Ecogenetics

Ecogenetics

Genetic predisposition to the toxic effects of chemicals

Editor-in-Chief Philippe Grandjean

Editorial Board Dinko Kello
Gunter Röhrborn
Stanislaw Tarkowski

Springer-Science+Business Media, B.V.

First edition 1991

Originally published by Chapman & Hall in 1991
Softcover reprint of the hardcover 1st edition 1991

ISBN 978-94-010-5383-9

British Library Cataloguing in Publication Data
Ecogenetics : genetic predisposition to the toxic effects of chemicals.
1. Man. Diseases. Genetic factors
I. Grandjean, Philippe II. World Health Organization
Regional Office for Europe
616.042
ISBN 978-94-010-5383-9 ISBN 978-94-011-3136-0 (eBook)
DOI 10.1007/978-94-011-3136-0

Library of Congress Cataloguing-in-Publication Data
Ecogenetics : genetic predisposition to the toxic effects of chemicals : proceedings of a WHO meeting held in Krefeld, Federal Republic of Germany, 17–20 October 1989 / editor-in-chief, Philippe Grandjean, editorial board, Dinko Kello . . . [et al.].–1st ed.
p. cm.
Based on papers presented at the Consultation on Genetic Predisposition to Toxic Effects of Chemicals.
Includes bibliographical references and index.
ISBN 978-94-010-5383-9
1. Genetic toxicology–Congresses. 2. Disease susceptibility–Genetic aspects–Congresses. 3. Human genetics—Variation—Congresses. I. Grandjean, Philippe. II. Kello, D. III. World Health Organization. Regional Office for Europe. IV. Consultation on Genetic Predisposition to Toxic Effects of Chemicals (1989 : Krefeld, Germany)
RA1224.3.E26 1991
616'.042–dc20 90–20973
CIP

Contents

Contributors

D. P. Agarwal
Institute of Human Genetics, University of Hamburg, Hamburg, Germany

Z. Bankowski
Council for International Organizations of Medical Sciences (CIOMS), Geneva, Switzerland

K. Berg
Institute of Medical Genetics, University of Oslo, Oslo, Norway

M. Blanusa
Institute for Medical Research and Occupational Health, University of Zagreb, Zagreb, Croatia, Yugoslavia

N. P. Bochkov
National Research Centre of Medical Genetics, Academy of Medical Sciences, Moscow, USSR

E. J. Calabrese
Northeast Regional Environmental Public Health Center, Environmental Health Sciences Program, School of Public Health, University of Massachusetts, Amherst, MA, USA

D. Djuric
Institute of Occupational and Radiological Health, Belgrade, Yugoslavia

H. W. Goedde
Institute of Human Genetics, University of Hamburg, Hamburg, Germany

P. Grandjean
Department of Environmental Medicine, Odense University, Odense, Denmark

J. Z. Hanke
Institute of Occupational Medicine, Lodz, Poland

K. Hemminki
Department of Industrial Hygiene and Toxicology, Institute of Occupational Health, Helsinki, Finland, and Center for Nutrition and Topicology, Karolinska Institute, Huddinge, Sweden.

W. Kalow
Department of Pharmacology, University of Toronto, Toronto, Ontario, Canada

F. Kaloyanova
Department of Toxicology, Institute of Hygiene and Occupational Health, Sofia, Bulgaria

B. Kargacin
Institute for Medical Research and Occupational Health, University of Zagreb, Zagreb, Croatia, Yugoslavia

D. Kello
Environment and Health Service, WHO Regional Office for Europe, Copenhagen, Denmark

S. R. Kleeberger
Department of Environmental Health Sciences, The Johns Hopkins Medical Institutions, Baltimore, MD, USA

K. Kostial
Institute for Medical Research and Occupational Health, University of Zagreb, Zagreb, Croatia, Yugoslavia

R. C. Levitt
Department of Anesthesiology and Critical Care Medicine, The Johns Hopkins Medical Institutions, Baltimore, MD, USA

T. Maljkovic
Institute for Medical Research and Occupational Health, University of Zagreb, Zagreb, Croatia, Yugoslavia

M. A. Mehlman
Department of Environmental and Community Medicine, Robert Wood Johnson Medical School, Piscataway, NJ, USA

M. Piasek
Institute for Medical Research and Occupational Health, University of Zagreb, Zagreb, Croatia, Yugoslavia

B. Raikhlin
The Israel National Poison Information Center, Haifa, Israel

E. Reiner
Institute for Medical Research and Occupational Health, University of Zagreb, Zagreb, Croatia, Yugoslavia

H. W. Rüdiger
Unit of Toxicogenetics, Institute of Occupational Health, University of Hamburg, Hamburg, Germany

V. Simeon
Institute for Medical Research and Occupational Health, University of Zagreb, Zagreb, Croatia, Yugoslavia

C. Sonich-Mullin
Environmental Criteria and Assessment Office, US Environmental Protection Agency, Cincinnati, OH, USA

D. T. Teitelbaum
Medical Toxicology Partnership, Denver, CO, USA

N. V. Titenko
National Research Centre of Medical Genetics, Academy of Medical Sciences, Moscow, USSR

Preface

Protection of human health by prevention of environmental hazards has become a major concern in all European countries. This concern is reflected in the regional strategy of Health for All by the Year 2000, as developed in 1984, and in the 1989 European Charter on Environment and Health. However, despite considerable progress, most efforts are related to single sources of exposure and to a small number of 'priority' chemicals.

In protecting the population against chemical hazards, attention must also be paid to the fact that subgroups are particularly susceptible to the adverse effects of certain chemicals. This hypersusceptibility may be genetically determined. Such individual predisposition to toxic effects must be taken into account in the efforts to prevent adverse effects of environmental exposures to toxic chemicals.

To help address these problems the Regional Office, in co-operation with the Institute of Human Genetics and Anthropology of the University of Düsseldorf and with financial support from the Federal Ministry for the Environment, Nature Conservation and Nuclear Safety of the Federal Republic of Germany, organized a consultation in Krefeld in October 1989. The meeting was attended by 17 experts from 12 countries, three representatives of the host government, two representatives of other organizations and five staff members from the WHO Regional Office for Europe and WHO Headquarters. The list of participants is included as an appendix to the present volume.

The fact that the present book was delivered within a short time is a tribute to all concerned, and I would like to thank all authors for their dedication and enthusiasm, and in particular the Editor-in-Chief.

J. E. Asvall
Regional Director, WHO Regional Office for Europe
Copenhagen, Denmark

Foreword

Genetic variability between humans is immense. The DNA present in the haploid human genome corresponds to about 3×10^6 average-size gene sequences, thus perhaps leading to some 20 000 polymorphic proteins. Some of these interindividual differences are likely to affect the susceptibility to chemical exposures. However, according to available data, only about 50 polymorphic traits appear to be of significance in relation to environmental chemicals. Furthermore, the epidemiological evidence is very limited, and little has been done so far to provide special protective measures for the individuals concerned. Although research in human genetic is proceeding rapidly, two factors have contributed to the limited action with regard to hypersusceptibility to environmental chemicals.

First, in the setting of exposure limits, the intention is to protect all individuals exposed. The traditional approach has been to use a safety factor of ten to account for any physiological or other causes of increased susceptibility to the exposures. Although this safety factor appears to have been sufficient in many situations, little evidence is available to document its overall validity. Continued reliance on a global safety factor may not be warranted. Thus, for example, studies of acatalasaemia and other inherited enzyme deficiencies now suggest that genetic predisposition in some cases may require much larger safety factors.

Second, if hypersusceptible individuals cannot be protected by feasible exposure limits, an alternative approach is to identify the individuals concerned and protect them by other means. Thus, screening should then be considered. Although WHO has recommended general principles for employing such methods, particular ethical concerns may be raised in relation to genetic screening and the use of this information. These issues need to be resolved before such screening can be initiated.

Thus, genetic predispositions to effects of toxic chemicals raises some very thorny questions. The Consultation on Genetic Predisposition to Toxic Effects of Chemicals in Krefeld, Federal Republic of Germany, from 17–20 October 1989, was held to help answer these questions.

As will be seen from the contributions made by the experts to the Consultation and from the report, genetic predisposition is but one of numerous factors that may affect health response to toxic chemicals.

However, its importance in relation to other factors is still not yet sufficiently defined. Further research must be more strongly oriented towards studying clinical relevance, and must also be better linked with growing knowledge of genetic predisposition to selected common diseases.

S. Tarkowski
Director, Environment and Health Service
WHO Regional Office for Europe
Copenhagen, Denmark

PART ONE

Significance for Public Health and Research

1 Report on a WHO consultation

1.1 INTRODUCTION

Individual predisposition to toxic effects may be classified into three types:

1. factors that increase the concentration of the biologically active substance at the active site;
2. factors that augment the reaction of the chemical substance with target molecules in the body, thereby initiating the response; and
3. factors that promote the sequence of events between the initial reaction and the final manifestation of an adverse health effect.

The term hypersusceptibility is preferred over other terms, such as hypersensitivity or hyperreactivity. Thus, hypersusceptible individuals will experience an adverse health effect to one or more pollutants at significantly lower or shorter exposures than the general population because of one or more factors that predispose the individuals to the harmful effects.

The availability of increasing numbers of DNA probes, the polymerase chain reaction, and other technological advances have greatly improved the prospects for identification of hypersusceptible individuals. The public health and research implications of the new information in this area must therefore be considered.

The specific purpose of the meeting was to review existing knowledge about, and evaluate the significance of, genetic variation in relation to the adverse effects on health of exposure to chemicals, and to assess the need to develop criteria for the identification and protection of population groups with such predispositions. In addition, the meeting was expected to identify the need for further research.

The participants elected Professor Jeffrey R. Idle as Chairman, Professor Gunter Röhrborn as Vice-Chairman, and Professor Philippe Grandjean as Rapporteur. A large part of the meeting was reserved for discussions of specific problems in subgroups. The following report has been compiled by the Rapporteur and is based on contributions from these subgroups. The individual conclusions and recommendations have been agreed upon by the participants.

The report first outlines the scientific background as based on ecogenetic, epidemiological and experimental evidence. Gaps in knowledge are identified, and strategies are suggested for strengthened and co-ordinated research efforts. From current knowledge, the potential significance for public health is discussed, in particular the need for dissemination of information, discussion of ethical aspects, setting of appropriate exposure limits and utilization of screening procedures.

1.2 SCIENTIFIC BACKGROUND

1.2.1 Ecogenetic evidence

Genetic factors play a considerable role in the metabolism of chemicals in the human body. Phenylketonuria (PKU) is an extreme example of genetic predisposition to chemical exposure. PKU patients suffer from a deficiency of phenylalanine hydroxylase, and the presence of normal amounts of phenylalanine in the diet invariably leads to severe neurological symptoms. However, early detection of a high concentration of phenylalanine in plasma and subsequent treatment with a phenylalanine-free diet can ameliorate the neurological symptoms. In other inherited metabolic diseases, such as Wilson's disease and lactase deficiency, hypersusceptibility also occurs (to copper and lactose, respectively).

Genetic polymorphisms have been determined for numerous enzymes that metabolize foreign compounds. Many of these enzymes have functions with dual consequences; not only do they aid in detoxicating toxic compounds in the body, but often they also cause biological activation of chemicals, such as conversion of prodrugs into drugs or of procarcinogens into carcinogens. Thus, interindividual differences in enzymic activity can now explain the existence of subjects hypersusceptible with respect to certain chemicals. The frequency of such inherited hypersusceptibility may vary considerably between different human populations.

Recent advances have occurred in the understanding of the molecular biology and epidemiology of alpha$_1$-antitrypsin, glucose–6-phosphate dehydrogenase (G–6-PD), cytochrome P–450 isozymes, alcohol and aldehyde dehydrogenases, *N*-acetyltransferase and cholinesterase. More limited insight has been obtained for other enzymes, such as sulfoxidases, paraoxonase, epoxide hydrolase, glucoronyl transferases and glutathione transferases. Thus, genetic factors can affect the occurrence and severity of environmental disease due to single as well as mixtures of chemicals.

However, the main emphasis so far has been on pharmacogenetics rather than toxicogenetics.

Certain genetic enzyme variants will have a high impact on individual predisposition to development of toxic effects and of various diseases, such as cancer. In particular, genetic variants of one of the cytochromes (P450IID6) constitute a major risk factor for lung cancer in smokers, and the genetically controlled acetylation capacity is of significant importance for the occurrence of industrially related bladder cancer.

Genetic differences between individuals may lead not only to quantitative differences in response to drugs or other chemicals but also to qualitative ones. Thus, haemolysis may occur in exposed G–6-PD-deficient subjects, and allergic responses may be seen in other situations.

Wider significance of genetic predisposition is also suggested by other recent observations. Thus, in human cells from some individuals, increased formation of DNA adducts or of cytogenetic abnormalities (such as micronuclei) may occur after a standardized in vitro exposure. Furthermore, impairments of DNA repair processes have been detected (i.e., deficient O^6-methylguanine-DNA-methyltransferase). So far, the interindividual variation in concentrations of DNA adducts is about 10-fold at the same level of carcinogen exposure; with binding of carcinogens to proteins, the variation ranges from 2- to 5-fold for most exposures. Although the implications of these observations are still unclear, they tend to support the notion that a variety of mechanisms may be involved in genetic predisposition to cancer and other diseases.

1.2.2 Epidemiological evidence

Genetic variability in the metabolism of environmental chemicals is likely to affect the occurrence and severity of several environmental diseases. However, most epidemiological data on exposure-related effects indicate average disease risks in a particular population. In studies of occupational exposures, the gainfully employed individuals constitute a highly selected part of the population. Even so, groups of workers are likely to exhibit a range of susceptibilities to toxic chemicals, although predisposed subgroups are rarely identified in such studies. So far, little consideration has been given to methodological approaches to identification of subgroups with increased susceptibility.

Other variables, such as age, nutrition, pre-existing disease and life-style, also contribute to differences in susceptibility among individuals. The contributions of all of these factors may make it extremely difficult to sort out the importance of a single factor, such as genetic predisposition.

A further obstacle is the clinical reliance on morphologic rather than mechanistic diagnostic criteria; this tradition results in limitations in the understanding of genetic factors influencing the development of environmental disease.

Despite these difficulties, ample evidence has been collected from family studies and other epidemiological studies that genetic differences contribute in important ways to the variability of responses to environmental agents. The diseases involved include both stochastic and nonstochastic disorders. Conversely, in some instances, genetic factors confer special resistance to some environmental influences.

Thus, genetic factors play a role in every common cancer. However, the magnitude of cancer risk associated with individual genes cannot yet be determined. Also, the pathogenesis and manifestations of genetically-related allergic and immunologic disease are still poorly understood.

With regard to coronary heart disease, lipoprotein Lp(a) is an important risk factor. A high level of Lp(a) in serum accounts for about 25% of myocardial infarction cases in males below 60 years of age. However, the significance of Lp(a) and other inherited risk factors in relation to chemical exposures is unknown at present.

Descriptive epidemiological data have suggested that the risk of developing a disease in many cases is related to genetic factors. For example, statistical associations have been documented between random qualitative genetic markers (such as blood groups) and diseases (such as cancer) that may also be related to chemical exposures. Such associations should stimulate search for specific underlying mechanisms.

Very limited epidemiological information is available on specific interaction between genetic factors and chemical exposures with regard to disease risk. However, DNA probes will soon be available so that specific genetic markers can be used in epidemiological studies. Also, improved exposure data may be collected simultaneously, because many chemical carcinogens react covalently with DNA; these chemicals also bind to proteins, and the concentrations of such adducts may become a useful marker in epidemiological studies.

1.2.3 Experimental models

Pathogenic mechanisms and biochemical markers, as well as dose–response relationships, are difficult to identify and characterize in humans. Thus, predictive experimental models of genetic predisposition are needed to enable qualitative risk assessment.

Toxicological testing of new drugs and other chemicals has led to the development of numerous animal models, but most of them do not account or control for genetic background. These studies generally use highly inbred strains that are reared under identical environmental conditions with the exception of the chemical exposure being tested. Such inbred strains are derived from a single breeding pair in the twentieth or a subsequent generation of brother-sister mating, or its equivalent; all members are genetically identical and homozygous at all loci. Alternatively, F_1 hybrids may be used, i.e. the first generation cross of two inbred strains; the animals are therefore isogenic, though heterozygous at all loci at which the parental strains differ.

While such testing methodologies may be useful for other reasons, they are likely to yield results that reflect the homogeneity of animals exposed. Consequently, variability in the population is usually markedly diminished, and dose-response relationships tend to be steeper. Therefore, knowledge about potential genetic contributions to individual susceptibility to effects of chemicals has generally been unobtainable.

However, comparison of effects in different strains of inbred animals may permit investigators to address this important factor. For example, such studies have focused on immune responses and disease resistance related to lipopolysaccharide-induced interleukin–1 production, development of amyloidosis and infections with *Listeria, Salmonella, Leishmania* and *Mycobacterium.*

Generally, an initial step to understanding genetic control of a biological process would be to screen a series of inbred strains for their specific responses to the agent under study. Interstrain variation in effects of xenobiotics indicates that the response(s) may be under genetic control. A difference between strains may be due to a single gene or a complex genetic relationship among multiple genes. Interstrain differences cannot be used alone to describe the nature of a genetic relationship, and more sophisticated genetic analyses, including breeding studies, must be utilized.

Two other types of animals are used for this purpose: (1) animals from a genetically heterogeneous or nonisogenic group, such as an outbred stock, F_2 hybrid or genetically heterogeneous colony; and (2) animals expressing a major gene mutation. An outbred stock is usually maintained as a closed colony, either with random mating or a mating system designed to minimize inbreeding.

One of the great strengths of an experimental model would be the ability to test hypotheses about mechanisms of the response under study, so that a primary defect leading to the disorder may be identified. Single-gene models are particularly useful because the techniques of cosegregation and candidate gene analysis may be utilized.

Cosegregation analysis has been utilized to develop an understanding of the molecular mechanisms of diverse phenomena, such as toxic chemical depression of the bone marrow, variation in metabolism of chemical carcinogens and susceptibility to development of tumours.

Candidate gene analysis has been used extensively to investigate whether the locus of the disorder in question 'maps' to the very region of a candidate gene that has previously been characterized. A successful match on the chromosomal map will support the significance of the gene for development of the disorder. If a match is not detected, for example if the two loci are on different chromosomes, then the candidate gene can be excluded as a potential mechanism for the disorder in question.

Studies in experimental animals have also been useful in elucidating metabolic patterns for toxic chemicals. Such studies suggest that the number of DNA-binding products (adducts) in a sensitive tissue will correlate with the tumour frequency. Although the biological significance of different types of DNA adducts is unclear, the validity of these techniques should be further explored.

1.3 SIGNIFICANCE FOR RESEARCH

1.3.1 Research needs

The number of recognized human genetic variants of xenobiotic-metabolizing enzymes is extremely small when compared to the number of such variants that must be postulated on the basis of general genetic variability. This discrepancy is primarily due to the fact that the technical capability to perform the biochemical measurements has only very recently become available.

Major differences between human populations in xenobiotic-metabolizing capacities have been established, mostly in terms of different frequencies of defined enzymic abnormalities or of gene frequencies. Less well defined are the quantitative or multifactorial differences of drug-metabolizing capacity between human groups, whether defined by age, gender, occupation, ethnicity or other criteria.

For instance, little is known about the occurrence of deficiencies of epoxide hydrolase, and enzyme that detoxicates reactive intermediates, or of the consequences of the well-established variation of glutathione transferases, especially in relation to exposures to xenobiotics.

Special attention should therefore be given to epidemiological studies on exposed groups of workers and the analysis of statistics on morbidity with emphasis on hypersusceptible individuals. The 'healthy worker effect' must be recognized as an obstacle in applying the results obtained from occupational studies to hypersusceptible individuals in the general population. Mathematical methods need to be further refined to enable more sensitive analysis of frequency distribution curves to identify multimodalities.

Very limited information is available on specific genes responsible for genetic predisposition or resistance in relation to environmental disease. So far, only phenotypic studies have been performed. The associations of cancer and respective genotypes are still unexplored since adequate gene probes are not yet generally available. The development of specific and selective DNA-probes for polymorphisms of interest is now possible and should be explored. For example, the genetic progress in relation to atherosclerotic disease indicates that the prospects will rapidly improve in the near term.

Animal models for genetic predisposition need to be further developed and characterized. Also, closer ties need to be developed between animal models and related human studies. In particular, an important objective of animal models will be to establish linkage associations of genetically determined susceptibility factors with phenotypic markers which are analogous in the mouse and human genomes.

Single-gene (Mendelian) models are particularly useful because of the approximately 80% homology between human and mouse genomes. Both cosegregation and candidate gene analysis are promising techniques.

At present we cannot quantify an environmental risk by using any of the genetic parameters available. The same is true for adduct determinations. Thus, the risk associated with, for example, 3 fmol benzo(a)pyrene adducts per μg DNA (for example, caused by exposure to polycyclic aromatic hydrocarbons) is totally unknown.

We have established several promising risk parameters, but we cannot use them for an individual risk assessment. This situation is comparable to the problem encountered 20 years ago with serum cholesterol levels and the association with coronary heart disease; the connection was clearly established in principle, but an individual's risk could not be quantified on the basis of a cholesterol determination.

1.3.2 Research strategies

Scientific progress in this area mainly rests on the collection of relevant epidemiological data. Thus, first of all, population studies are urgently needed to determine the prevalence of genetic markers associated with predisposition to adverse effects of environmental exposures. Furthermore, extensive studies must be conducted to quantitate the magnitude of risks associated with the genetic factors for individuals exposed to specific environmental chemicals.

In this regard, large prospective studies will be required. Appropriate cohorts for the study of such factors should be identified and established as soon as possible so that adequate studies can be initiated. Specific studies should be performed in the occupational environment to permit the assessment of the health consequences of genetic hypersusceptibility to adverse environmental exposures.

Cohort studies must be initiated as soon as possible to obtain results within a reasonable time. The total cohort must be large enough to allow the collection of valid data also for disorders that are not among the most frequent ones. The cohort should be used in common by several groups and for the testing of various risk parameters, because the proper selection of a cohort and the follow-up of the cohort members are extremely expensive.

Proper stratification and selection should be performed with regard to gender, age, ethnicity and exposure to reduce the variance. Special attention should be paid to the problem of confounding factors (such as other exposures, life-style and social demographic factors) and the impact of a bias due to the exclusion criteria used.

These considerations imply that only a large international co-operative study would satisfy the requirements. Alternatively, an attempt could be made to join already existing cohort studies if the original purpose can be modified to fulfil the specific needs to elucidate toxicogenetic risks.

Epidemiological research is often exceptionally expensive, and this is particularly true in this case because of the need for sophisticated determinations (genotyping and phenotyping). Therefore, co-ordination of the research is necessary to minimize costs and increase the efficiency.

Genetic differences may be identified by analysis of human cells, tissues or body fluids for particular functional differences (such as enzyme activity) or structural differences (e.g. in DNA or proteins). Clinical and laboratory studies should be expanded to identify new genetic markers through the exploration and use of innovative screening methodologies. Laboratory studies should also be undertaken to develop interlaboratory standardization procedures for identification and quantitation of genetic markers and their effects. Distinction should be made between population studies of gene frequencies and screening studies aimed at spotting affected individuals for the purpose of medical intervention; the methods used in such studies may be of different precision and therefore of different cost.

Further development of methods for determining DNA and protein adducts should be encouraged. Such methods should be used to measure specific DNA and protein adducts as well as other biological indicators in the same individuals. Follow-up studies in populations of high cancer risk, such as patients undergoing chemotherapy, are needed to assess to what extent adduct levels predict cancer risk.

Because of the research implications of these modern analytical techniques, whole blood should be stored from people in cohort studies for future analyses, such as candidate gene studies. DNA in whole blood will be stable in the frozen state. Thus, in a follow-up study, when a sufficient number of individuals have contracted a disease thought to be caused by a particular exposure, and candidate genes are available for study, DNA should be prepared and analyzed (for restriction fragment length polymorphism, deletions, duplications) in affected individuals and matched controls. In this way, predisposing as well as protective genotypes should be identifiable. This banking approach could be included in many cohort studies without adding much to the expenses.

In view of the significant prospects for this research and the well-established differences of gene frequencies in different populations, WHO should assist the establishment of procedures so that appropriate specimens of human tissues (autopsy samples, blood specimens, etc.) can be collected in different countries and stored. Such specimen banks will allow large-scale studies of enzyme variations in individuals and populations. The comparative investigations should take place in a single reference laboratory or in a series of laboratories with established common standards of measurement.

The cohort studies should be accompanied by proper family studies to clarify modes of inheritance.

A successful outcome depends on a crucial international collaboration between the disciplines of human genetics and occupational/environmental medicine. Multicentre activities will be needed for the efforts to identify genetic predisposition and to determine its health significance. WHO and national authorities should therefore undertake steps to stimulate, fund and co-ordinate the research in toxicogenetics.

1.4 SIGNIFICANCE FOR PUBLIC HEALTH

1.4.1 Dissemination of information

Public information in relation to genetic predisposition and its potential public health significance is crucial for the optimal application of scientific insights and technological advances.

Such information may have consequences for individual choices, e.g. of nutrition, stimulant use and occupation. More specifically, an individual or a group of individuals must be properly informed if recognized as being genetically at high risk of developing adverse effects from a certain chemical exposure. If screening for genetic predisposition is carried out, counselling should be considered before the testing is performed.

General recognition of the role of genetic factors in the occurrence of environmentally induced disease is of great importance for understanding the complex web of the multifactorial aetiologies. Although the degree to which supporting data are available is highly variable, acquisition of such information is likely to be an invaluable tool for education, prevention, health care and research.

In the education of physicians and other medical personnel, the curricula should therefore incorporate the influence of genetic factors on the susceptibility to adverse effects of environmental exposures.

Occupational health services should be made aware of the potential hypersusceptibility of some workers. Approaches to prevent such individuals from being exposed must be developed from a prudent judgment on the basis of current evidence.

In research and scientific documentation, the field of genetic predisposition suffers from difficulties in information retrieval. A new scientific database should be developed, or improvements should be made in existing databases to collect and disseminate information on ecogenetics. In particular, improvement in indexing (with adequate key words) would be desirable to simplify searches for relevant information.

1.4.2 Ethical aspects

The occurrence of genetic predisposition may be of considerable importance because of the potentials for reducing the risk of disease in employees or consumers. Early diagnosis would potentially give carriers of the gene an opportunity to plan their lives to reduce the impact of the hypersusceptibility. Failure to detect and warn such individuals at risk could seem culpable. Predictive testing also promises benefits to the fortunate ones found not to be gene carriers.

However, testing for genetic predisposition may infringe upon individual freedom. Also, it may lead to violations of principles of fairness due to discrimination or stigmatization. Thus, the result of a predictive screening test will inevitably be of interest to others besides the person at risk, and this fact may lead to ethical conflicts.

While the public health context must be appreciated, a balance should be sought between the interests of the individual and those of society.

If screening for genetic predisposition is indicated, the procedures should follow ethical principles similar to those used in research on human subjects, as formulated in the *Helsinki Declaration* (World Medical Association) and the proposed *Guidelines for Research Involving Human Subjects* (Council for International Organizations for Medical Sciences and WHO).

Screening of human beings may be justified only if there is clear scientific evidence of the existence of genetic predisposition to toxic effects of a particular chemical. Restrictions must apply with regard to access and use of this information so that the interests of the individual are properly protected.

Only screening tests of high diagnostic validity should be used and procedures for follow-up detection of false positives should be considered. The magnitude of risk due to the predisposition should be considered when determining the specific screening procedures.

To avoid misuse of screening data, a set of guidelines should be developed in accordance with ethical principles used in research involving human subjects, i.e. autonomy, confidentiality and equity.

Thus, as in other fields of genetic counselling and diagnosis, testing for predisposition of individuals or groups of individuals should be voluntary, and the results should be handled confidentially. The collection of data through genetic screening programmes may otherwise result in discriminatory practices and other negative social consequences. Thus, the information must be handled with extraordinary security and discretion. The patient should be properly informed of the meaning of the result, but the utilization of the information given should again be determined on a voluntary basis.

Likewise, research in genetic predisposition and development of potential screening methods may result in ethical concerns. The principles generally used in medical research should also be applied in relation to these types of research.

Questions on ethical problems cannot be settled by experts. A public discourse on the potential conflicts in relation to screening and related questions would seem highly pertinent. Conflicts concerning ethical issues can only be resolved in an open, informed forum.

1.4.3 Exposure limits

Primary preventive measures should be initiated to improve the health and quality of life, when genetic predisposition to environmental exposures has been identified. Such measures include the determination of safe limits of exposure to environmental chemicals.

When such permissible exposure limits are set, both experimental and epidemiological evidence are taken into account. However, current regulatory approaches in this area are inadequate to protect hypersusceptible subgroups.

Depending upon the strength of the epidemiological and experimental data available, uncertainty factors (or safety factors) are employed. A variety of uncertainties are considered by this approach, including the possible existence of particularly susceptible members of the population. Generally, a 10-fold uncertainty factor is used to account for likely variability in susceptibility among individuals. Although based on qualitative considerations only, this approach is not yet known to have led to catastrophic adverse effects in human exposures to a wide range of chemicals. However, no one knows how protective the uncertainty factor really is, either in general or in specific cases.

In an attempt to ascertain the extent to which a 10-fold dose reduction protects hypersusceptible members of a population, log probit slopes from experimental animal studies have been examined. The majority of the slopes (92%) for acute effects were greater than 3; a 10-fold dose reduction decreased the average response by at least three standard deviations, i.e. protecting at least 99.9% of the animals. For the remaining 8% of the chemicals considered, a 10-fold reduction in doses would not be sufficient to protect hypersusceptible animals. Although these calculations do not specifically address genetic predisposition, they illustrate the fact that a 10-fold uncertainty factor is unlikely to have a general validity.

The 10-fold factor may be even less adequate to protect humans, because human populations are not as homogeneous as experimental animals, and because other environmental exposures, health status, sex, age and nutritional status may play an additional role. Furthermore, pharmacogenetic evidence indicates that hypersusceptible individuals may in some cases need a safety factor of more than two orders of magnitude. Thus, the selection of the number 10 has not been based on any quantitative assessment of interindividual variability in susceptibility. On the contrary, the proportion of the population that is hypersusceptible is generally regarded too small to merit separate consideration. Although current legislation on occupational exposures aims at protecting hypersusceptible individuals, i.e. mostly young workers and pregnant women, genetic predisposition is not taken into account. However, for a population exposed to multiple chemicals, the proportion that can be considered hypersusceptible may be substantial; various subgroups will be hypersusceptible to at least one of the chemicals.

In the setting of environmental and occupational limits for exposures to hazardous chemicals, international or national legislation should aim at protecting all individuals, including those who may be hypersusceptible due to genetic predisposition. Thus, current regulatory practices need to be improved in this regard. To the extent that uncertainty factors are needed, they must be carefully considered on a chemical-specific basis, depending on the strength of the data base. In some instances, uncertainty factors much above 10 may be justified.

1.4.4 Screening

Large-scale population-based genetic screening programmes will potentially provide information that can improve health and the quality of life for individuals as well as for populations. Such screening studies should be carried out with appropriate protection of the rights and privileges of all persons concerned. For example, although perhaps not directly related to chemical exposures, screening for lipoprotein Lp(a) would be of considerable value, because individuals with a high risk of coronary heart disease could be identified.

In some countries, pre-employment examinations are used to obtain information on, for example, prior allergies or disease. However, screening for genetic hypersusceptibility has not been conducted on a large scale. Thus, in 1982, limited genetic screening was being performed at a small number of industrial facilities in the USA. However, the participants were unaware of any cases of current genetic testing in the workplace.

Routine medical surveillance may be utilized to advise and protect those workers who may be hypersusceptible. Such surveillance should in particular cover the early period of employment to identify unexpected hypersusceptibility so that protective measures may be instituted as early as possible.

Although many methods offer some promise for potential detection of individuals hypersusceptible to toxic effects of chemicals, they are not fully validated. Thus, there is insufficient scientific justification to recommend their use in obligatory pre-employment screening procedures. This conclusion also applies to tests for G–6-PD deficiency, sickle-cell trait, thalassaemia, and alpha$_1$-antitrypsin deficiency.

With regard to screening for such types of hypersusceptibility in the future, the genetic nature of the trait tested must be clearly understood and this trait must be a major factor in the development of a disease of public health importance. In addition, the potential for subsequent action must be considered; only when a screening result can lead to potential beneficial efforts should the screening test be considered.

Thus, the following are prerequisites for screening for genetic predisposition:

1. high prevalence of the genetic condition;
2. a health risk in hypersusceptible persons considerably higher than in other persons similarly exposed;
3. the availability of a sensitive and specific screening test.

In addition to specific tests for genetic predisposition, determination of DNA and protein adducts may be used to identify individuals who, due to their exposures and/or genetic predisposition, are at an increased risk of cancer.

1.5 CONCLUSIONS AND RECOMMENDATIONS

1.5.1 Conclusions

1. The existence of genetic polymorphisms for the metabolism of foreign compounds has been amply demonstrated. The resulting interindividual differences in enzyme activity, which may be several hundred-fold in

magnitude, explain the existence of subjects who are hypersusceptible to the effects of both drugs and industrial chemicals.

2. Genetic factors may either predispose people to suffer from the adverse effects of environmental exposure or protect them against such effects. The diseases that are influenced by such genetic factors include atherosclerosis, various malignancies, and allergic and other immunological disorders, as well as specific intoxications.

3. The quantitative measurement of chemical adducts with DNA and proteins is now possible; such methods indicate exposure to certain chemicals. These methods also show the range of variability between individuals in metabolic activation and in DNA repair.

4. Animal models of genetic susceptibility to environmental chemicals are useful tools to illuminate the mechanisms involved, but are of limited use in predicting genetic predisposition in humans.

5. Although taking account of age, gender, nutrition, lifestyle and existing disease, both epidemiological studies and regulatory practices have not sufficiently addressed the significance of genetically hypersusceptible subgroups.

6. Individuals with an identified predisposition to suffer from the adverse effects of environmental chemicals require the special attention of public and occupational health services.

7. When adequate and sufficient information exists, detection of the genetic characteristics that influence people's susceptibility to environmentally induced diseases may serve to guide intervention, and improve the health and quality of life of the affected individual or population.

8. The current regulatory practice for setting health-related environmental or occupational limits for hazardous chemicals does not specifically take account of groups in the population who are hypersusceptible owing to genetic differences. Existing regulation in most countries is restricted to protecting groups such as young people and women.

9. The conventional application of uncertainty/safety factors when standards are set cannot fully guarantee the protection of hypersusceptible individuals. Current knowledge indicates that people may exhibit more than a hundred-fold variability in their susceptibility to chemicals.

10. Validated screening procedures exist for the detection of individuals hypersusceptible to cardiovascular disease and those hypersensitive to certain chemicals.

11. The detection of variations in individual response to chemical exposure raises serious ethical dilemmas, such as equal rights to enjoy

opportunities in society, equal rights in free enterprises, and rights to privacy, individual freedom and good health.

1.5.2 Recommendations

1. Primary preventive measures, such as the control of hazardous exposures, should be preferred as a means of protecting individuals with a genetic predisposition to react to them.

2. The limits for environmental and occupational exposure to hazardous chemicals should protect all individuals, including those who may be hypersusceptible. Procedures to achieve this goal need to be developed.

3. Screening for any genetic predisposition should be performed only when the following prerequisites are fulfilled:
- the prevalence of the genetic predisposition is significant;
- susceptible persons have a significantly higher health risk than other persons similarly exposed; and
- a sensitive and specific screening test is available.

4. In co-operation with national authorities, WHO should disseminate information to help clarify the ethical aspects of screening for genetic predisposition to disease. National authorities should stimulate ethical inquiry into related preventive strategies, and guidelines for screening procedures should be prepared, reflecting the basic human rights to good health, confidentiality, autonomy and equity.

5. WHO should explore the possibilities of improving current databases on scientific information by including index terms that reflect ecogenetic subjects.

6. WHO should encourage the development of curricula for health personnel and public health information for the layman on the genetic determinants of individual susceptibility to biologically active chemicals.

7. Research should take new directions, including:
- large prospective cohort studies to address genetic host susceptibility, particularly for polymorphic genes that govern the variable activation and detoxification of chemicals;
- family studies to characterize the mode of inheritance of genetic predisposition;
- improvement of methods to determine macromolecular adducts and other biological markers in the same individual; these methods should be used to detect and characterize hypersusceptible subgroups;
- animal models need to be developed and refined to obtain a better understanding of genetic susceptibility to chemical exposure;

- basic research in statistical modelling and risk assessment methods to assess the health risks to hypersusceptible individuals.

8. Research programmes need better co-ordination, particularly between different countries, to improve the use of resources and the harmonization of procedures, and to establish joint cell and tissue banks from well-defined populations. WHO should explore its role in this area. New sources of funding should be made available by governments to ensure the satisfactory conclusion of such studies.

PART TWO

Perspectives of Genetic Predisposition

2 Genetic predisposition to occupationally-related diseases: current status and future directions

E. J. Calabrese

Northeast Regional Environmental Public Health Center, Environmental Health Sciences Program, School of Public Health, University of Massachusetts, Amherst, MA, USA

2.1 SUMMARY

This report evaluates the general issue of whether susceptibility to occupationally-related diseases may be enhanced by an individual's genetic constitution and to what extent any such genetic factors may be identified (as in the form of biochemical markers) in ways which are applicable to widespread screening within the workforce. Because of the potentially broad scope of the topic, eight specific genetic conditions or disease entities are selected for critical evaluation: sickle cell trait/disease, glucose–6-phosphate dehydrogenase (G–6-PD) deficiency, thalassemia, serum $alpha_1$-antitrypsin (SAT) deficiency, cytochrome P–450 enzymes (e.g. acetylator phenotype), chemically induced allergic hypersensitivities and diabetes.

2.2 HISTORICAL PERSPECTIVE

While the study of genetic susceptibility to environmental/occupational agents has a long history, this field entered the so-called modern era in the late 1950s with the report of Motulsky (1957) who found that specific types of drug toxicity were caused by genetic disorders affecting drug-metabolizing enzymes. These findings led Vogel (1959) to coin the term 'pharmacogenetics' to describe the interaction of heredity and environ-

Table 2.1 Identification and quantification of genetic factors affecting susceptibility to environmental agents* (from Calabrese, 1984)

High-risk groups	*Estimated occurrence*	*Environmental agents to which group is (may be) at increased risk*
RBC condition		
G6PD deficiency	American black males 16%; Mediterranean Jewish males 11%; Greeks 1%–2%; Sardinians 1%–8%	Environmental oxidants such as ozone, nitrogen dioxide, and chlorite
Sickle cell trait	7%–13% of American blacks are heterozygotes	Aromatic amino and nitro compounds; carbon monoxide, cyanide
The thalassemias	Alpha: 4%–5% in Americans of Italian and Greek descent; Beta: 2%–7% American blacks and 2%–3% American Greeks	Lead; benzene
NADH dehydrogenase deficiency (MetHb reductase deficiency)	Estimated 1% of population are heterozygotes	MetHb-forming substances
Catalase		
Hypocatalasemia	About 2% of US population based on Swiss gene frequency	Ozone; radiation
Acatalasemia	1/10 000–20 000 of US population based on Swiss gene frequency	
Low SOD activity	Frequency of genetic variants in population 1 to 2/10 000; normal population exhibits unimodal distribution; persons at low end of distribution may be at increased risk	Wide variety of environmental oxidants; paraquat; radiation; ozone
ALA dehydratase deficiency	Unknown, but thought to be rare	Lead
Hb M	Unknown, but rare	Carbon monoxide
Erythrocyte porphyria	1.5/100 000 in Sweden, Denmark, Ireland, West Australia; 3/1000 in South African whites; rare in blacks	Chloroquine; hexachlorobenzene; lead; various drugs, including barbiturates; sulfonamides, others

High-risk groups	*Estimated occurrence*	*Environmental agents to which group is (may be) at increased risk*
GSH-Px deficiency	Rare	Environmental oxidants
GSH deficiency	Rare	Environmental oxidants
Liver metabolism		
Defect in glucoronidation		
Gilbert's syndrome	6% of normal, healthy adult population	Wide variety of xenobiotics including polychlorinated biphenyls
Crigler-Najjar syndrome	Few persons live to adulthood	
Defect in sulfation	Unknown	Wide variety of xenobiotics; best association is with tyramine-containing foods
Acetylation phenotype, slow vs fast	Slow: 50% white; 50% blacks; 10% Japanese	Aromatic amine-induced cancer; numerous drugs, e.g. isoniazid and hepatitis
	Fast: 50% white; 50% blacks; 90% Japanese	
Gout	0.27%–0.3% prevalence in US and Europe	Lead
Oxidation centre defects	9% of British whites; 8% of Nigerians; 6% Ghanians; 1% Saudi and Egyptians are poor oxidizers	Numerous xenobiotics requiring oxidative metabolism for detoxification
OCT deficiency	Unknown, but thought to be rare	Insect repellant (DET)
Paraxonase variant	25%–30% of population	Parathion
Rhodanese variant	Unknown	Cyanide
Sulfite oxidase deficiency heterozygotes	Unknown	Sulfite, bisulfite, sulfur dioxide
Inadequate carbon disulfide metabolism	Upward of 30%–40%	Carbon disulfide
Alcohol dehydrogenase variant	5% English; 20% Swiss; 70% Japanese	Metabolize (e.g., ethanol) more quickly than normal
Wilson's disease	Homozygous 1/100 000 while the heterozygote may approach 1/500	Copper, vanadium

High-risk groups	*Estimated occurrence*	*Environmental agents to which group is (may be) at increased risk*
Serum variants		
Albumin variants	Less then 1/1000 in Europeans, much higher frequency in North American and Mexican Indians	Unknown
Pseudocholinesterase variants	Highly sensitive homozygous and heterozygous persons of European ancestry have combined frequency of about 1/1250; moderately sensitive genotypic variants of European ancestry have frequency of 1/15 000	Organophosphate and carbamate insecticides; muscle relaxant drugs
lgA deficiency	1/400–600	Respiratory irritants
Serum α-antitrypsin deficiency – heterozygote	Approximately 4%–9% of Northern Europeans	Respiratory irritants, smoking
Cardiovascular conditions		
Hyperlipidemia	5% exhibit polygenic hypercholersterolemia	Carbon disulfide, carbon monoxide, smoking
Homocystinuria heterozygotes	1%–2% in population	Carbon disulfide, carbon monoxide, smoking
Respiratory condition		
High aryl hydrocarbon hydroxylase inducibility	Approximately 9%	Polycyclic aromatic hydorcarbons
CF heterozygotes	4% of white	Respiratory irritants
Renal diseases		
Renal nephropathy of diabetes	5% of adult population	Heavy metals; organic solvents
Cystinosis	Most common among whites of European ancestry; unknown frequency	Heavy metals
Cystinuria	1/200–250 persons, although asymptomatic are heterozygote carriers; 1/20 000–100 000 are homozygous	Heavy metals
Tyrosinemia	Frequency in general population unknown; frequency of heterozygous carriers in Chicoutimi region of northern Quebec is carrier for every 20–31 persons	Heavy metals

High-risk groups	*Estimated occurrence*	*Environmental agents to which group is (may be) at increased risk*
Immunological disorders		
Immunodeficiency diseases – hypoagammaglobulinemia lgA deficiency (see serum variants)	Very rare	Viruses; bacteria
Immunological hypersensitivity	2% of some worker populations	Isocyanates
Dermatological conditions		
Alibinism:		
Tyrosinase-positive	1/14 000 blacks; 1/60 000 whites; very high frequency in American Indians	Ultraviolet radiation
Tyrosinase-negative	1/34 000–36 000 blacks and whites; 1/10 000–15 000 in Ireland	Ultraviolet radiation
PKU	1/80 in US is a carrier; 1/25 000 has disease	Ultraviolet radiation
Ocular disease		
Glaucoma	5% at high risk; 30% at moderate risk	Glucocorticoid drugs
Leber's optic atrophy	Thought to be rare	Cyanide
Thyroid disorders		
Nontasters of PTC	30% Europeans; 10% Chinese; 3% Africans	Giotrogenic substances
DNA repair and chromosome instability disease		
XP	1/250 000 homozygous recessive; 2–4/1000 heterozygote carriers; higher incidence in North American Arabs	Wide range of mutagens and/or carcinogens
AT	2.5/100 000 homozygous recessive; 1% heterozygote carriers	
FA	1/300 heterozygote carriers; 3/1 000 000 homozygous recessives	
Bloom's syndrome	Unknown frequency	
Down's syndrome	1/636–1/1000 births	

*Abbreviations used are: G6PD, glucose-6-phosphate dehydrogenase; NADH, nicotinamide adenine dinucleotide (reduced form); SOD, superoxide dismutase; ALA, aminolevulinic acid; GSH, reduced gluthathione; Px, peroxidase; OCT, ornithine carbamoyl transferase; lgA, immunoglobulin A; PKU, phenylketonuria; PTL, phenylthiourea; XP, xeroderma pigmentosum; AT, ataxia telangiectasia; FA, Fanconi's syndrome.

ment in the pathogenesis of disease. Over the next 20 years considerable evidence was accumulated to establish a wide range of genetic factors that affect xenobiotics (Jensen, 1962; Kalow, 1962; Stokinger, 1962; Zavon, 1962; Brieger,1963; Meier, 1963; Mountain, 1963; Price-Evans, 1963; Stokinger and Mountain, 1963). Concurrent with these cited efforts was the publication for the first time in 1960, of the now classic reference text, *The Metabolic Basis of Inherited Disease*, which offered biochemical foundations to describe the occurrence of genetic factors influencing the metabolism of a wide variety of xenobiotics.

By the end of the 1960s Stokinger *et al.* (1968) sought to close the gap between the pharmacogenetic perspective with its emphasis on drugs and the environmental perspective, by showing its application to the field of industrial hygiene. Further development of the environmental associations with genetic factors led Brewer in 1971 to propose the term 'ecogenetics', a phrase he argued that encompassed the field dealing with genetic predispositions to both drugs and any type of environmental agent.

The field of ecogenetics has continued to provide significant insight into the causes of interindividual variation in response to drugs and pollutant exposures. One particular current concern has dealt with the capacity of ecogenetic research to assist in the understanding of interindividual variation in susceptibility to carcinogenic agents. Recent years have seen a strong emphasis directed toward the legal, social and moral implications of how the knowledge of genetically based differential susceptibility to environmental agents should be used (Mason *et al.*, 1986; Lappe, 1988).

Amongst the several hundred human genetic diseases that have been identified (Stanbury *et al.*, 1983) approximately 50 have been established as having the potential to enhance an individual's susceptibility to the toxic/carcinogenic effects of environmental agents (Table 2.1) (Calabrese, 1984, 1986).

This report is intended to provide an update of earlier reports (Calabrese 1984, 1986) which critically reviewed information on selected genetic conditions which are either known or reasonably theorized as enhancing susceptibility to environmental agents. From a long list of genetic conditions that may enhance susceptibility to occupational toxins (Table 2.1), eight previously reviewed conditions and four new genetic factors were selected for evaluation based on the adequacy of the database, nature of the toxic endpoint and the frequency of the trait in the population. In addition, this report will provide an identification of changes in technology that have or are likely to have a significant impact on future progress in this field. The report also contains a series of institutional recommendations on how the field of ecogenetics could be markedly advanced and a judgement on the database to support screening for genetic predispositions to occupationally-induced diseases.

2.3 SPECIFIC CONDITIONS AFFECTING SUSCEPTIBILITY TO ENVIRONMENTAL CONTAMINANTS

2.3.1 Erythrocyte G–6-PD deficiency

It has been shown that G–6-PD-deficient individuals are at enhanced haemolytic risk from a number of oxidant drugs and industrial chemicals. Many substances commonly used in industry are known to cause haemolytic changes and have also been speculated to present an enhanced risk to G–6-PD deficients. However, the number of specific industrial substances for which reasonably definitive proof exists that G–6-PD deficients are at greater haemolytic risk than normal is limited to certain aromatic amino and nitro compounds (for example naphthalene and TNT respectively). No adequate quantitative risk assessment of the haemolytic actions of these substances on G–6-PD deficients has been published, although it would appear that this information may be available for the aromatic nitro and amino compounds. There is a growing body of in vitro evidence strongly implicating the enhanced susceptibility of G–6-PD-deficient erythrocytes to a wide variety of industrial and environmental oxidants (Calabrese, 1984, 1986). Unfortunately, it has not been possible to relate such in vitro exposures to ambient or industrial exposures. There is also a general lack of epidemiologic studies designed to assess the impact of industrial and environmental oxidants on G–6-PD deficients. Furthermore, since haemolytic crises may be precipitated in deficients by numerous commonly used medications and bacterial infections, it would be of great interest to assess the effect of these factors in combination with industrial haemolytic compounds known to present at least an in vitro enhanced risk to G–6-PD-deficient erythrocytes.

Concerted attempts have been made to assess and develop predictive models for G–6-PD deficiency (Horton and Calabrese, 1986). However, no model satisfying the needs for both qualitative and quantitative extrapolative predictions has emerged. For example, despite its use to simulate the response of human G–6-PD deficients to oxidant stressor agents (Amoruso, 1985), the C57 L/J mouse strain is not a suitable predictive model since it displays relatively higher erythrocyte G–6-PD activity than human deficient erythrocytes. In fact, it simulates G–6-PD activity of normal human erythrocytes rather than of deficient erythrocytes.

In contrast to the C57 L/J mouse model, Dorset sheep have low erythrocyte G–6-PD activity and have been proposed as a possible predictive model (Calabrese, 1984). The sheep erythrocytes frequently display enhanced susceptibility to a range of oxidant stressor agents as do G–6-PD-deficient human erythrocytes. Furthermore, the predictive utility of the sheep model has been enhanced by the application of an S–9 bioactivation system (Bloom *et al.*, 1983; Horton and Calabrese, 1986a). Despite the obvious attractiveness of the sheep model it also has limitations, although not as serious as the C57 L/J mouse model. For example, there are interspecies differences in the activity levels of other antioxidant enzymes. In addition, Dorset sheep display GSH (reduced glutathione) levels characteristic of normal humans. They also usually display oxidant stress with marked increase in MetHb (methaemoglobin) levels while human deficients respond with more modest MetHb increases but more striking GSH depletion (Horton and Calabrese, 1986b). In summary therefore, the sheep model has qualitative predictive potential when properly interpreted.

Another approach involving the use of the heterologous erythrocyte model (Castro *et al.*, 1973), in which human deficient cells are injected into immunocompromised rats, has not been sufficiently developed to be of predictive use (Horton and Calabrese, 1986a). Finally, the use of steroid hormones to inhibit tissue G–6-PD activity has been attempted; the steroid treatments diminish hepatic but not red cell G–6-PD activity in vivo (Calabrese *et al.*, 1987a, 1987b), thereby precluding the development of an erythrocyte G–6-PD-deficient model by this approach.

2.3.2 Sickle cell anaemia/trait

Whether sickle cell trait carriers are at increased risk from the challenges of rigorous training at high altitudes remains unresolved. Limited evidence suggests that possession of sickle cell trait is a contributing factor in the reported cases of injury and/or death, but that other as yet undetermined factors must also be present. No experimental or epidemiological evidence has confirmed the 1973 hypothesis of Stokinger and Scheel that persons with sickle cell trait are at increased risk from anaemia producers, MetHb formers and blood oxygen tension reducers.

2.3.3 Thalassemia

While limited clinical observations have suggested that persons with thalassemia may be at increased risk to benzene- and lead-induced hematotoxicity, the data are insufficient to be convincing. Nevertheless, suggested clinical studies support the need for epidemiological investigations and/or development of a predictive animal model to test the association between thalassemia and lead- and/or benzene-induced haematotoxicity (Calabrese, 1984; Carta *et al.*, 1987).

2.3.4 Acetylation phenotype

A growing body of epidemiological studies suggests that slow acetylation is a predisposing factor for the recurrence of arylamine-induced bladder cancer (Weber, 1984, 1987). Further research has indicated that susceptibility to bladder cancer is not equal among those generally grouped as 'slow acetylators', but that susceptibility is greater as the acetylation capacity decreases (Cartwright *et al.*, 1982). Consequently, future studies should consider acetylation-ratio subgroups as well as the typical fast and slow acetylator phenotype comparisons.

When one considers that 50% of the North American Caucasian and Black populations are slow acetylators, the sheer magnitude of those at potentially increased risk is striking. The present situation is that the theory is well founded in cancer research with a variety of animal models. However, epidemiological studies of industrial bladder cancer populations (Omenn, 1986) are needed, to investigate a possible correlation between the risk of bladder cancer and the ability to acetylate arylamines.

While there is incentive to initiate epidemiological investigations in this area, it must be emphasized that considerable progress has been made in the evaluation of animal models of highly variable hepatic acetylator status (Hultin, 1983; Weber *et al.*, 1984; Hultin and Weber, 1986; Mattano and Weber, 1987). These collective findings of Weber and colleagues have established the occurrence of hereditary polymorphisms in the *N*-acetylation of arylamine and hydrazine drugs and various carcinogens in rabbits, mice, hamsters and rats. The results of Weber (1987) indicate that genetic analysis of the acetylator tract in rabbits, mice and hamsters is regulated as an autosomal trait that may be classified as either rapid or slow in a similar manner to that performed in humans. It is expected that this range of potential animal models will offer considerable opportunities for biomedical researchers in the evaluation of specific hypotheses.

2.3.5 Cytochrome P–450 enzymes

(a) Introduction

Since many, if not most, environmental toxins require bioactivation to highly reactive intermediate forms capable of causing a wide array of toxic cellular/molecular lesions, the importance of cytochrome P–450 enzymes in the development of environmentally-induced disease is theoretically profound. The cytochrome P–450 enzymes are involved in the oxidative metabolism of not only an enormous range of endogenous molecules such as steroids, prostaglandins and biogenic amines, but also innumerable drugs, chemical carcinogens, mutagens and environmental contaminants. Given this critical role of cytochrome P–450 enzymes in the operation of Phase 1 metabolism of xenobiotics of environmental concern, it is not surprising that genetic variation in P–450 enzyme activities and their relative capacities for induction are likely to affect markedly susceptibility to an enormous array of environmental contaminants. Consequently, the cytochrome P–450 enzymes are given a special distinction in this analysis of genetic factors affecting susceptibility to environmental agents. The field of P–450 molecular biology, according to Nebert and Gonzalez (1987), has 'literally exploded' in the current decade with advances occurring in the multiple facets that surround P–450 research, such as chemicals that induce P–450 activities (Nebert *et al.*, 1981; Waterman and Estabrook, 1983), the association of P–450 with cancer (Pelkonen and Nebert, 1982; Conney, 1982; Wolf, 1986; Gonzalez *et al.*, 1987), genetic differences in P–450 expression (Nebert *et al.*, 1982) and others.

(b) Aryl hydrocarbon hydroxylase

Calabrese (1984, 1986) devoted considerable attention to genetic susceptibility to cigarette-smoking-related lung cancer and in particular to the hypothesis that susceptibility to polycyclic aromatic hydrocarbon (PAH)-induced lung cancer can be predicted by differential ability to induce the enzyme aryl hydrocarbon hydroxylase (AHH). This enzyme, which is found in most mammalian tissues, catalyses the initial step in the metabolism of benzo(a)pyrene (BP) and other PAHs. Since AHH is an inducible enzyme, it displays increased activity after administration of a variety of agents such as PAHs, various drugs, steroids and insecticides. Aryl hydrocarbon hydroxylase is believed to play a critical role in the bioactivation of PAHs such as BP by metabolizing them to epoxides. The epoxide may bind to DNA and other macromolecules or it may be converted to other products such as phenols and dihydrodiols or be conjugated to GSH. Diol-epoxides are believed to be the principal mutagenic metabolites of PAHs. Benza(a)pyrene diol-epoxide binding to DNA and other macromolecules is believed to be the initial cause of malignant transformation in cells. Thus, PAH metabolism via AHH results both in detoxification for excretion and in activation to more highly mutagenic and carcinogenic intermediates (Calabrese, 1984).

The key feature in the AHH-inducibility lung cancer hypothesis is the assumption that individuals can be accurately phenotyped by measuring AHH induction in mitogen-stimulated lymphocytes and that these levels are positively correlated with lung AHH levels. Unfortunately, research with lymphocytes from lung cancer patients has produced highly variable results, most likely because of poor survival, reduced protein synthesis and reduced mitogen response when these cells are grown in culture (see Calabrese, 1984 for a review).

Due to the experimental limitations in the testing and validation of the AHH-inducibility lung cancer hypothesis, considerable efforts were directed toward the identification of new genetic markers, ones that are easily determined and which, of course, are reproducible. This need coincided with the emergence of the relationship of debrisoquine hydroxylation status and susceptibility to environmentally-induced cancers (Calabrese, 1984). In fact, Islam *et al.* (1980) suggested that environmental carcinogens such as aflatoxin B_1, BP and dimethylnitrosamine, which are oxidized to chemically reactive metabolites in tissues, may have greatly differing effects in individuals according to their hydroxylation status, with rapid oxidizers being at greatest risk.

While it was not known at that time if debrisoquine hydroxylase was directly involved in the bioactivation of chemicals which are carcinogenic for the lung, Ayesh *et al.* (1984) hypothesized that

> its gene may flank others with oncogenic potential, recessives for which will almost certainly be recessives for debrisoquine hydroxylation. Such genes might be structural or regulatory genes for cytochrome P–450, such as the Ah locus or indeed genes not involved in chemical carcinogen metabolism, for example, cellular oncogenes . . . The metabolism oxidation phenotypes serve as genetic markers for differential lung cancer susceptibility.

In the late 1980s considerable progress was made concerning the family of P–450 enzymes, progress that also clarified the relationship of AHH to debrisoquine hydroxylase. In brief, the field of P–450 molecular biology showed enormous expansion in the last decade. For example, the first two P–450 3′ cDNA probes were reported in 1980 and the first P–450 full length cDNA sequence in 1983. By 1986, there were 67 complete P–450 cDNA or protein sequences in eight eukaryotic species and one prokaryote. It is now known that there are ten P–450 gene families with eight existing in mammals.

Based on cDNA nucleotide sequencing, coherent theories of P–450 gene evolution have emerged (Nebert and Gonzalez, 1987). In 1987 Nebert *et al.* proposed a standardized P–450 gene nomenclature involving the use of Roman numerals and capital letters that facilitates the matching of previously characterized P–450 proteins with the newly defined gene families and subfamilies. The mammalian P–450 families are designated I to VIII. For example, the P–450 I family in rats, rabbits, mice and humans has two genes (P_1 and P_3) that are inducible by PAHs and TCDD. In essence the human P_1 enzyme is what has been called AHH and is associated with an increased risk from cigarette-smoking-induced bronchogenic carcinoma. In contrast to P_1, P_3 is highly correlated with cancer induced by 2-acetylaminofluorene and aminobiphenyls in animal models (Kamataki *et al.*, 1983). The P–450 II gene family was initially referred to as the phenobarbital-inducible family. This gene family is now organized to comprise five subfamilies A-E with subfamily IID represented by debrisoquine 4-hydroxylase (P–450IID1). Within the IIE subfamily are at least two P–450 proteins that are induced by ethanol, immidazole, acetone, trichloroethylene and pyrazole.

According to Nebert and Gonzalez (1987), the P–450 activity showing the highest induction includes the oxidation of aniline, alcohols and nitrosamines. These authors anticipate the initiation of clinical studies attempting to correlate restriction fragment length polymorphism (RFLPs) patterns of these genes with individual risk of cancer. With respect to the P–450 III gene family, macrolide antibiotics such as triacetyloleandromycin and griseofluvin induce at least one gene of this family. Nebert and Gonzalez (1987) suggest that the results of clinical studies of the human orthologues of these P–450 III genes could be useful in the prediction of macrolide antibiotic response to or toxic effects of these agents.

(c) Debrisoquine metabolism defect(s)

Convincing evidence indicates that genetic factors affect the ability to oxidize molecules of dissimilar structure such as alicyclic, aliphatic and aromatic carbon centres. Of considerable relevance in this regard is that many known animal model/human mutagens and/or carcinogens require an initial bioactivation step via an oxidative process. To what extent humans differ in their ability to bioactivate toxic and/or carcinogenic compounds may contribute significantly towards explaining variation in population responses to such agents. A notable finding in this area was provided by Ayesh *et al.* (1984) who reported an association between extensive debrisoquine hydroxylation and the prevalence of bronchogenic carcinoma in smokers. Based on these findings it has been theorized that metabolism oxidation phenotypes such as debrisoquine hydroxylation serve as genetic markers for differential lung cancer susceptibility (Ayesh *et al.*, 1984). In addition, the number of potentially affected people is considerable, with 5–10% exhibiting a debrisoquine metabolism deficiency.

Prior to 1988 it was believed that poor metabolizers originated from a single mutation. However,in 1988 Gonzalez *et al.* revealed at least four mutations which all led to the poor metabolizer phenotype. Furthermore, these mutations occur not within the coding exons of the gene, but in the introns, thereby leading to a series of incorrectly formed messenger

RNAs that are not able to yield immunodetectable protein. These defects are not only unusual but are hard to detect and necessitate the sequencing of the complete normal human P450db1 gene, including introns, exons and flanking regions, before the source of faulty splicing of pre-messenger RNA in poor hydroxylators can be understood. In practical terms this greater complexity is expected to lead to a delay in the development of new approaches for definitive phenotyping and detection of poor metabolizers using techniques that do not involve drug metabolism. According to Idle (1988), if a single mutation were involved it would most likely have led to the relatively rapid development of a linked restriction fragment length polymorphism analysis of DNA obtained from a small blood sample. Despite the greater complexity, Idle expects that the molecular characterization of the debrisoquine hydroxylation polymorphism by Gonzalez *et al.* (1988) will eventually enhance the capacity to understand the relationships between environmental agents and genetic susceptibility to disease.

In a parallel research effort investigators have attempted to identify and/or develop a more effective marker of hepatic pharmacogenetic oxidative xenobiotic metabolism than debrisoquine. However, these efforts have often failed because of the absence of an effective endogenous marker substance and the lack of an obvious structure-activity relationship within a particular family of substrates. Despite the fact that a number of drugs have been proposed, a general marker for the assessment of the debrisoquine type drug hydroxylation polymorphism has not been able to satisfy the three principal criteria of an ideal phenotype test drug: (a) specificity, (b) general availability and (c) innocuousness. However, in 1986 Kupper *et al.* presented data to indicate that the widely employed drug dextromethorphan may adequately satisfy these requirements. Evidence was presented that dextromethorphan *O*-demethylation is catalysed by the debrisoquine-type cytochrome P–450 isozyme. This common genetic control of debrisoquine and dextromethorphan metabolism led the authors to recommend that this drug may be employed as a safe and innocuous substitute for debrisoquine for future phenotyping in the field of human pharmacogenetic oxidative drug metabolism. In practical terms the debrisoquine administration may be replaced by the dextromethorphan test procedure without a change in the experimental protocol. Thus, it is expected that dextromethorphan may become widely used for world-wide phenotyping without regard for age or ethnic origin.

Table 2.2 Summary of recent studies concerning susceptibility of serum alpha$_1$-antitrypsin (SAT) heterozygotes to the development of chronic obstructive pulmonary disease

Authors	*Purpose of study*	*Study design factors*	*Findings/comment*
Beckman *et al.* (1984)	Assessed relationship of the SAT-deficient allele to COPD	This was a limited study of heterozygotes (n = 13) in which pulmonary function was evaluated	A high proportion of those studied displayed respiratory symptoms
Menkes *et al.* (1984)	This was a genetic epidemiology study assessing multiple risk factors on obstructive airway pulmonary disease	Involved a cross-sectional study of 2539 non-patient adults	Potential risk factors were significantly associated with airway obstruction; age and smoking were most important; of secondary importance was protease inhibitor phenotype
Eriksson *et al.* (1985)	Assessed the role of smoking and heterozygote status as risk factors in the pathogens of emphysema	The study involved assessing the responses in 83 men selected randomly from a larger population evaluated 6 years earlier. There were: 18 nonsmokers (NS), with normal SAT (N) 13 smokers (S) with (N) 18 nonsmokers (NS) heteroxygous for SAT (H) 14 S-H	The H-NS did not differ from N-NS; but H-S displayed a greater decrease in $FEV_{1.0}$ than H-NS. Despite the decline in lung functions, there was no increased prevalence of clinical obstructive disease
Aderele *et al.* (1985)	Assessed whether there is a higher prevalence of SAT deficiency in asthmatic children	This study compared asthmatic (n = 55) with nonasthmatic (n = 55) children for SAT levels	The data indicated that SAT levels were significantly lower in the asthmatics; however, there was no evidence of emphysema in the children

Authors	*Purpose of study*	*Study design factors*	*Findings/comment*
Khoury *et al.* (1985)	This study examined the contribution of environmental and genetic risk factors in COPD	Using a familiar aggregation methodology, the authors assessed 325 first-degree relatives and 56 spouses of 150 COPD patients. These were compared with 222 first-degree relatives and 56 spouses of 107 nonpulmonary controls for the prevalence of airway obstruction and chronic bronchitis	While smoking (personal and parental) and SAT status were related to the endpoints of concern, other undefined factors also contributed to the indicators of COPD
Horn *et al.* (1986)	This study assessed whether there is an association between the heterozygous SAT state and respiratory functions	Using a case control methodology 28 heterozygote men were matched for age, years of employment and smoking status with those who were SAT normal. All had regular exposure to high levels of grain dust. Pulmonary functions tests, chest X-rays and skin tests were performed	The heterozygote had consistently poorer mean results for $FEV_{1.0}$, $FEV_{1.0}/FVC$, MMFR, and Vmax50; also, the heterozygotes had 3× as many individuals with abnormal chest X-rays; the results suggested that heterozygotes are at greater risk of COPD than those normal for SAT
Christensson *et al.* (1986)	This study assessed various lung function parameters in four groups: NS-N, S-N, H-NS, H-S	(See purpose of study)	The only significantly affected parameters were the volume of trapped gas (VIG) and this occurred only after bronchodilation. The authors suggested that VTG after bronchodiation may be a sign of alveolar abnormality preceding the development of lung emphysema

2.3.6 Serum $alpha_1$ antitrypsin (SAT) deficiency

Research from animal and human studies strongly supports the hypothesis that an intermediate deficiency of SAT is a contributing factor in the development of emphysema. However, it is now being recognized that emphysema is a multicausal disease and that the heterozygote state, by itself, is not a major predisposing factor in the development of the disease. However, in combination with other predisposing factors such as high leukocyte protease, the heterozygote condition may be at an enhanced risk of developing emphysema. The most reasonable position at the current time would be to evaluate the relative contribution of such multiple variables in the development of emphysema. If the above associations prove true and are amenable to widespread screening, then it is likely that screening would involve an assessment of multiple contributory factors (Calabrese, 1984, 1986).

Since my 1984 report (Calabrese 1984) there have been approximately seven studies of a clinical or epidemiological nature that have attempted to assess whether heterozygote carriers of SAT deficiency display enhanced risk of developing COPD. Despite the fact that these studies have somewhat different goals, and involved a variety of different research teams and different populations, the consistent conclusion is that the heterozygote displays a greater inclination for biological endpoints indicative of COPD (Table 2.2). Of particular note is that predisposition for the development of COPD occurs even in the absence of occupational and lifestyle (e.g. smoking) risk factors (Horn *et al.*, 1986), but that the risk is further exacerbated by smoking (Eriksson *et al.*, 1985) and occupational risk factors (e.g. grain dust) (Horn *et al.*, 1986). Furthermore, in a genetic epidemiology study by Khoury *et al.* (1985) other non-identified factors contributed to airway obstruction and chronic bronchitis after adjustments for personal (direct) and parental (secondary) smoking and SAT phenotype. While these studies have looked at the issue of SAT heterozygotes and the risk of COPD via different perspectives, they have essentially reiterated the same conclusion. However, despite the inherent quality and extensive efforts in these reports, none of these studies offers a significant or quantum advance on where the field was in the earlier 1980s. Unfortunately they tended to concentrate on the debate about whether or not the heterozygous status was an important risk factor in COPD.

The key factor that has not been appreciated by research since the early 1980s is the need to identify other independent genetic risk factors, such as macrophage protease activity, and link these factors to an understanding of the disease process itself. Such an approach, being both biologically sound and oriented towards the disease process, increases the likelihood of quantifying the role of various genetic factors affecting the development of COPD.

One area that showed strong biomedical potential but which has not been sufficiently exploited to date is the use of an SAT-deficient rat model. Blackwood *et al.* (1984) were able to render rats deficient in SAT by administration of galactosamine and they successfully applied it to the evaluation of pulmonary changes induced by bacterial endotoxin. It appears that this model has the potential to be effectively exploited for studies involving interaction with other COPD risk factors.

2.3.7 Chemically-induced hypersensitivities (allergic)

While many of the genetic conditions identified here as possibly enhancing susceptibility to chemical agents act via a metabolic condition of the host such as G–6-PD deficiency, many idiosyncratic toxicities have an immunological aetiology or hypersensitivity (allergic) basis (Pohl *et al.* 1988). Various authors have estimated that the hypersensitivity class of idiosyncratic reactions may account for between 3 and 25% of all drug reactions (Patterson and Anderson, 1982; De Weck, 1983; Goldstein and Patterson, 1984; Mathews, 1984). The severity of drug allergies is quite diverse, ranging from slight inconvenience to life-threatening events.

In order for a xenobiotic agent to be recognized as nonself and to induce an immune response it is generally thought that the agent must have a minimum molecular weight of about 1000 Daltons (De Weck, 1983). Since most drugs and industrial agents are smaller than this, it is generally assumed that they need to become covalently bound to an endogenous carrier macromolecule to form a drug carrier conjugate in order to be able to interact with the immune system (Parker, 1982; De Weck, 1983; Schneider, 1983; Park *et al.*, 1987). Drugs and industrial agents may form drug carrier conjugates after having been bioactivated through biotransformation processes. Examples of industrial agents that form complexes with macromolecules following metabolic activation include dinitrofluorobenzene, phthalic anhydrides, diisocyanates, diisothiocyanates, formaldehyde and ethylene oxide (Karol *et al.*, 1978; De Weck, 1983; Grammer *et al.*, 1984; Mathews, 1984; Patterson *et al.*, 1986; Park *et al.*, 1987).

Despite all the circumstantial evidence of this causal connection, only in the instance of halothane hepatitis have the identity of the bound metabolite and the characterization of the endogenous carrier molecules been ascertained (Pohl *et al.*, 1988). Once the drug (or industrial agent) carrier conjugate is formed, it may then act as an immunogen and cause a specific humoral response or a cellular T-lymphocyte response, or both responses. The type of toxicity caused by a specific xenobiotic is believed to be a function of the relative distribution of its drug carrier conjugates.

As is the case with many other diseases, immunologically-mediated hypersensitivities also involve complex webs of causation with numerous processes (e.g. bioactivation of the xenobiotic, differential half-life of the carrier conjugate, differential activation of the immune system) that are independently regulated. Such a complex of genetic factors affecting the occurrence of hypersensitivity responses creates the basis for the relatively low frequency in the population, thereby making it difficult to study this phenomenon in humans.

According to Park *et al.* (1987) no definitive study has been published on the relationship between interindividual variation in drug metabolism capacity and the incidence of drug hypersensitivity. Progress has been less than ideal for a number of reasons.

1. Well characterized allergic drug reactions are typically rare and, therefore, prospective metabolic immunotoxicity studies are not easily initiated.
2. Retrospective drug metabolism investigations are not likely as a result of the risk associated with rechallenging with the allergen.
3. Chemically reactive metabolites are hard to quantify since they are usually chemically unstable and frequently minor ($<5\%$) products.
4. Animal models have as yet been of extremely limited extrapolative value in most instances of suspected drug allergy.

What is the future? Despite the impressive challenges noted above and taking their imposed limitations into account, Pohl *et al.* (1988) indicate that it is now technically feasible to identify and characterize drug carrier conjugates involved in some drug hypersensitivities. Upon purification, the drug carrier conjugate can serve as antigen for identifying sensitized persons who have had a hypersensitizing response to a xenobiotic. These authors argue that such knowledge could be employed to prevent not only other allergic reactions to the xenobiotic agent but possibly also cross-reactions to other structurally related agents.

2.4 NEW GENETIC PREDISPOSITION AREAS FOR CONSIDERATION

2.4.1 Methylation

Methyl conjugation is a significant pathway in the biotransformation of many xenobiotics. However, despite the fact that it has been widely cited in texts on biochemical detoxification, it has not been emphasized to the same extent as have glucuronidation and GSH conjugation. However, in the mid to late 1970s sensitive assays were developed that were able to determine the activities of methyltransferase enzymes in blood (Raymond and Weinshilboum, 1975; Pazmino and Weinshilboum, 1978; Weinshilboum *et al.*, 1978) and provided the opportunity to assess the regulation of these enzymes in humans. Over the past decade, it has been established that genetic determinants play a significant role in controlling individual variation in human methyl conjugating enzymes (Weinshilboum and Raymond,1977; Weinshilboum and Sladek, 1980; Keith *et al.*, 1983). Four methyl conjugating enzymes have been reported for which human polymorphisms have been found. These enzymes are catechol-*O*-methyltransferase (COMT), thiopurine methyltransferase (TPMT), thiol methyltransferase (TMT), and histamine *N*-methyltransferase (HNMT). Red blood cell COMT and TPMT activities are controlled by a common genetic polymorphism. Similarly, red blood cell (RBC) TMT and HNMT have high heritability; in these two cases the mode and mechanism of inheritance remain to be clarified.

While extensive research efforts have elucidated the presence, activities and modes of inheritance at least in part of the above mentioned methyl conjugating enzymes in human RBC, the key issue is whether RBC enzyme activity levels reflect the same enzyme activity in organs such as the liver and kidneys which play critical roles in xenobiotic metabolism. The answer emerging from the work of Weinshilboum and colleagues is that it is indeed highly reflective of the levels in other tissues (kidney, lung, lymphocyte) and may in fact be of good surrogate predictive value (Sladek-Chelgren and Weinshilboum, 1981; Weinshilboum and Raymond, 1977).

Inherited variations in RBC COMT activity reflect not only variation of the enzyme activity in other tissues but also individual differences in the metabolism of certain drugs. In the case of COMT, a significant positive correlation exists between the genetically determined RBC COMT activity and the proportion of L-dopa converted to 3-*O*-methyldopa in treated patients with Parkinson's disease (Reilly *et al.*, 1980). In the case of TPMT, it catalyses the the *S*-methylation of thiopurine drugs such as mercaptopurine (6-MP) and azathioprine that are widely employed for their cytotoxic and immunosuppressive activity. 6-Thioguanine nucleotides (6-TGN) are the principal metabolites of 6-MP and azathioprine in humans. Red blood cell 6-TGN levels are directly associated with risk for the development of leucopenia in patients with thiopurine drugs. Lennard *et al.* (1987) established that an inverse association exists between RBC TPMT activity and RBC levels of 6-TGN. Their findings indicated that low TPMT activity might be a risk factor for elevated 6-TGN concentrations and TPMT activity might make it possible to predict this risk factor for the development of thiopurine drug toxicity (Reilly *et al.*, 1980; Campbell *et al.*, 1984; Lennard *et al.*, 1986, 1987).

The approach of Weinshilboum and colleagues involved the measurement of enzyme activities in RBC and has shown that inherited variation in methylator status may help to account for variation in metabolism of and response to thiopurine and catechol drugs that undergo methyl conjugation. Of relevance to the present report is that Weinshilboum (1987) has hypothesized that inherited variations in methylation may also play a role in individual differences in susceptibility to toxicity after exposure in industry and in the environment to xenobiotic compounds that are capable of undergoing methylation.

The frequency of various phenotypic expressions of the polymorphisms of the various methyltransferase enzymes in humans has partially been defined. More specifically, for TPMT, one in 300 Caucasians are deficient, and 11% are heterozygote carriers of the deficiency (Lennard *et al.*, 1987), while for COMT, 25% are homozygous recessive and 50% are heterozygous carriers (Weinshilboum, 1984).

The pharmacogenetics of methyl conjugation and its environmental/occupational health implications were not presented in earlier reviews. However, in the current decade this topic has received considerable attention by Weinshilboum and colleagues at the Mayo Clinic. Most of the research has been directed toward establishing a sensitive RBC assay methodology that is applicable to a number of related methyl conjugating enzymes, associating RBC enzyme activity with tissue levels and establishing the mode of inheritance of these traits in humans. Little application has yet been directed toward the toxicological implications of these polymorphisms. However, low levels of TPMT constitute a risk factor for 6-TGN-induced toxicity, and indirect evidence suggests that the TPMT polymorphism is also a factor contributing to the occurrence of hypoprothrombinemia in patients treated with cephalosporin antibiotic drugs such as moxalactam that form heterocyclic sulfhydryl metabolites that undergo *S*-methylation catalysed by TPMT (Kerremans *et al.*, 1985). Thus, TPMT can catalyse the *S*-methylation of a variety of aromatic and heterocylic thiol compounds other than thiopurine (Woodson *et al.*, 1983; Ames *et al.*, 1986 Lennard *et al.*, 1987). Applications to the field of environmental health/industrial hygiene are likely to be markedly enhanced by the availability of inbred mice and recombinant inbred strains (Weinshilboum, 1987).

Background on methylation

Methylation is a two step conjugation process, initially involving the biosynthesis of *S*-adenosylmethionine which is the coenzyme donor and, secondly, the transfer of the methyl groups from the coenzyme to the acceptor molecule, mediated by specific transmethylases. Methylation has long been recognized as important for the metabolism of several endogenous chemicals, including the *N*-methylation of natural amines (e.g. norepinephrine, serotonin and histamine), the *O*-methylation of catecholamines and hydroxyindoles and the *S*-methylation of thiol compounds (de Bruin, 1976). There appears to be a multiplicity of mammalian transmethylases which are classifiable according to substrate specificity and location in multiple tissues.

Despite its role in the metabolism of numerous endogenous substrates, it has been generally accepted that only a few foreign compounds undergo biological methylation. For this reason, methylation has tended to be ignored in the environmental/occupational domain. Nevertheless, a number of environmentally important methylations do exist. For example, methylation has been shown to convert pyridine to *N*-methylpyridiniumhydroxide, and selenate to dimethylselenide. Methylation also occurs for arsenic, tellurium and mercury (de Bruin, 1976).

2.4.2 Sulfate conjugation

Another metabolic pathway not considered in earlier reviews was that of sulfate conjugation. However, varying levels of phenol sulfotransferase have been found in human platelets (Anderson *et al.*, 1981), and these variations reflect the variation in enzyme activity in other tissues. Variations in platelet activity of phenol sulfotransferase have been correlated with sulfate conjugation of acetaminophen (Reier and Weinshilboum, 1982) and O-methyldopamine (Campbell *et at.*, 1984). As in the case with methyl conjugation these examples show that regulation of drug metabolizing enzymes in human blood elements such as platelets may parallel their regulation in other organs. Such a quantitative association can provide a functional metabolic probe yielding valuable ecogenetic information in humans.

Despite the important initiatives of the above cited studies concerning sulfate conjugations, the database remains very limited and its role in accounting for differential susceptibility to xenobiotics is far from clear. Nevertheless, given the importance of sulfate conjugation in the overall context of conjugation strategies further evaluation of its quantitative relevance is warranted.

Background of sulfate conjugation

Sulfate conjugation plays an important role in the excretion of certain xenobiotics, especially those with phenolic groups. Sulfate conjugation is commonly involved in the metabolism of steroids and certain carbohydrates (de Bruin, 1976). Compounds commonly forming arylsulfates directly include a wide variety of monosubstituted phenols, such as chloro-, methyl-, amino-, acetamido- and nitrophenol. Of occupational relevance is that aniline, 1 or 2-naphthylamine and benzidine undergo *N*-sulfurylation yielding the corresponding sulfamates. This involves about 8% of the doses administered. Mono- and polyhalogenobenzenes are also metabolized by sulfate esterification via intermediary formation of the halogenated monophenols and catechols, but this reaction is less important than glucuronidation. Similarly, moderate sulfate conjugation has been reported for other agents, including monochloronitrobenzenes, aniline, polycyclic aromatic amines (e.g. 4-aminobiphenyl, 2-naphthylamine), polycyclic aromatic hydrocarbons (e.g. naphthalene, chlorinated naphthalenes, phenanthrene, pyrene), as well as amino azobenzenes.

2.4.3. Diabetes

The effects of pollutants on diabetics constitute an emerging area of environmental toxicology. The fact that kidney complications such as glomerulosclerosis are so common and serious in diabetics clearly underlines the need to assess the extent to which people with such diabetes-related pathology may be at increased risk to the wide range of renal toxins such as heavy metals, selected antibiotics, anesthetic gases and a variety of organic solvents such as 1,4-dioxane and related agents (Calabrese, 1984).

2.4.4 Genetic predisposition to viral agents

While this paper has focused on genetic predisposition to chemically-induced toxicities, genetic factors also affect susceptibility to micro-organisms such as bacteria, fungi and viruses (Calabrese, 1984). Of particular topical importance is a report by Eales *et al.* (1987) which demonstrated an association of different allelic forms of group specific component (Gc) with susceptibility to and clinical manifestations of human immunodeficiency virus infection. Group specific component is an α–2-globulin found both on cells and in serum and has three allelic forms based on electrophoretic analysis. It has been proposed that Gc may affect viral entry into the host cells (T cells and monocytes), the efficiency of this is related to the allelic form of Gc. It is interesting to note that the allele (Gc if) highly associated with susceptibility to the AIDS virus is found in 16% of Europeans and 58% of central Africans while the frequency of the allele (Gc) not associated with susceptibility to infectivity with the AIDS virus is 28% in Europeans and only 6% in Central Africans.

2.4.5 Multiple chemical exposures and genetic predisposition

Nearly all research on genetic predisposition to environmental agents has focused on single agents. However, it is beyond dispute that daily life results in multiple chemical exposures. Concern over multiple agent exposures has led many organizations (WHO, ACGIH, EPA, NAS) to establish theoretical frameworks with which to deal with this problem (Calabrese, 1991). None of these organizations has specifically addressed the issue nor the potential differential effects of multiple agent exposure in high risk groups including those with genetic predispositions to environmental disease. However, exciting research in the area of differential effects in slow and fast acetylators concerning the effect of ethanol ingestion on the metabolism of a variety of drugs illustrates this concept and also highlights differential response to multiple agents based on identifiable genetic factors (Weber, 1987). More specifically, ethanol consumption increases hepatic acetylation with the slow acetylator being markedly more affected than the fast acetylator.

2.5 DISCUSSION

The area of genetic predisposition to environmental agents is extremely broad. This makes it somewhat challenging to chart the progress on all fronts over the last half of the decade. During this time the field has been marked by 'spotty' advances. Little progress for example has been made in determining whether sickle-cell trait and thalassemia increase risk to specific environmental agents, while enhancement of our understanding of the relationship of G–6-PD deficiency and the risk of haemolytic responses has been limited by inadequate animal models and a paucity of epidemiological studies. As for SAT deficiency, interest has remained high and a number of significant studies have been published. Unfortunately, the efforts addressing SAT deficiency have not been coupled with other independent genetic conditions that may be predisposing factors, such as differential protease levels in macrophages. Notable exceptions appear to be in the area of cytochrome P–450 enzymes and acetylation status. The reason for progress in these areas may be related to strong investigator initiatives, but also to the high general interest in genetic determinants of cancer. This uneven level of research in the area of genetic predispositions is most likely a reflection of the lack of a long-term overall research agenda in the field of ecogenetics.

Even though there have been a number of excellent symposia on the topic of genetic factors which enhance susceptibility to environmental toxins, none of these meetings also included the development of a research needs consensus. Thus, the field in general has not had a research funding organization committed to a long-term research agenda in this area; there are no broad-based centres of excellence in this area nor any federal research agency devoting substantial resources to a co-ordinated development of the field of ecogenetics. What does exist are pockets of individual investigator interests which are dependent on the uncertainty of funding cycles. While such an approach had led to research advances it has understandably produced uneven advances at times unrelated to the relative importance of the area of study.

2.6 TECHNOLOGIES: PRESENT AND FUTURE

2.6.1 Gene transfer

The insertion of foreign genes precisely where desired into the recipient genome is likely to be a major factor in the field of ecogenetics. The field has been extremely limited by the number of quantitatively accurate predictive animal models available. This has been shown to be the case for G–6-PD deficiency, sickle cell trait/anaemia and various thalassemia disorders.

The principal challenge in gene transfer technology that needs to be overcome presently is the frequent lack of expression of the transferred genes. It has been speculated that this may result because the transferred genes do not carry all the regulatory sequences required for normal expression. This is leading the researchers in this area into what is referred to as targeted gene transfer. This involves fixing a gene in situ, thereby leaving its regulatory sequences intact. The targeted gene transfer approach has the capacity to introduce specific mutations into the model which would be sought for ecogenetic research as well as for repair of gene defects (Marx, 1988). Of relevance is that on 15 November 1988 the DuPont Company announced it will begin selling genetically engineered mice early in 1989. The first product is called 'OncoMice' because they carry human oncogenes. The strain of mice has the 'ras' oncogene, which is associated with a variety of human cancers, inserted into its genome. In addition, a tumour virus was also incorporated into their genetic makeup markedly enhancing susceptibility to cancer development (Gladwell, 1988).

2.6.2 Gene sequencing

Since the polymorphic expression of cytochrome P–450 genes has emerged as a central factor in explaining the occurrence of adverse drug side effects, as well as cancer and other toxicities, the continued generation of cDNA probes for the identification of these polymorphisms will likely prove to be of considerable value. Such probes will be of extreme importance in the clinical assessment of specific P–450 genes in predicting toxicity caused by xenobiotics since they could be used to phenotype clinical drug metabolizing capacity. The methodology makes use of the fact that even though DNA sequences at homologous loci are nearly identical from one copy to the next, sequence variations are frequently found. Improved techniques can uncover these variants in the DNA sequences and employ them as genetic markers. Consequently, the DNA sequencing technique can be used to track the inheritance of specific regions of the chromosomes.

Another common source of detectable sequence variation are regions with tandem repeats of short DNA sequences (Bell *et al.*, 1982; Jeffreys *et al.*, 1985a, b). The number of tandem repeats in the regions may vary from one chromosome to another. If a restriction fragment contains such a region, it will thus define numerous alleles. Kouri *et al.* (1986) indicate that these regions of tandem repeats are the best genetic markers thus far characterized and these markers have the potential to provide individual specific 'fingerprints' of human DNA (Jeffreys *et al.*, 1985b).

Restriction fragment length polymorphisms (RFLPs) have been applied to the challenge of phenotyping for capacity of drug metabolism. For example, White (1985) demonstrated that individuals who carried the disease congenital adrenal hyperplasia were defective in the gene coding for P–450c21. These data supported the hypothesis that this disease results from a deletion of a specific P–450 gene. Kouri *et al.* (1986) have indicated it is probable that RFLPs will be found associated with all p–450 genes because of the high level of diversity among humans at the level of the DNA sequence. They emphasized that activity-related RFLPs might not be located near the structural genes for P–450 but instead in close proximity to the regulatory genes (i.e. in the case of P_1–450 and P_3–450). Individuals are likely to have different drug metabolism phenotypes as a result of different regulator not structural genes (Pelkonen and Nebert, 1982; Kouri *et al.*, 1983).

2.6.3 Presence of microsomal preparations

In 1983 Bloom *et al.* pioneered the application of liver microsomes to test potentially haemolytic drugs in human erythrocytes with a G–6-PD deficiency. This procedure was soon applied by Calabrese *et al.* (1987a) to a large number of agents of environmental and occupational health concern in both sheep and human erythrocytes. This basic concept was applied by Spielberg (1984) in the study of cultured human lymphocytes with drugs in the presence of microsomal preparations. An advantage of this system according to Spielberg is that it offers an experimental approach to the pharmacogenetics of human idiosyncratic drug reactions that does not result in the exposure of patients to further xenobiotic-related risk. For example, he used lymphocytes from patients deficient in glutathione synthetase and established that such cells with diminished glutathione defences have enhanced susceptibility to metabolites of a variety of drugs. While this in vitro technique using a microsomal preparation is useful in ecogenetics research its applications are considerably broader.

2.6.4 Direct measurement of a drug-metabolizing enzyme activity

This approach has frequently taken advantage of the presence of enzyme activities in readily available peripheral tissues such as blood. Weinshilboum (1984) notes that enzyme activities have often been assessed in blood and as a surrogate for typically inaccessible tissues (such as lung, liver and kidney). However, this approach depends upon the assumption that enzyme activities in blood reflect either the regulation or properties of the enzymes in drug-metabolizing origins. It is critical that this assumption be carefully validated in order to develop a predictive system.

2.7 RESEARCH RECOMMENDATIONS

General institutional/programmatic recommendations

1. A committee should be formulated to create a long-term research agenda in the area of ecogenetics. It should consider both the role of epidemiological investigations and predictive animal models in the elucidation of the role of genetic factors affecting susceptibility to environmental agents.
2. A commitment should be made to the development of multiple research centres devoted to the study of specific aspects of the genetic predisposition challenge.
3. A commitment should be made to hold an annual scientific conference on recent findings and developments in the field.
4. A specific journal devoted to the publication of peer-reviewed papers should be created to enhance technical information transfer further; the journal would be an ideal vehicle to publish the conference proceedings.

2.8 GENERAL RESEARCH APPROACHES: IDENTIFYING AND OVERCOMING OBSTACLES

Several major challenges arise in the evaluation of genetic factors affecting susceptibility to toxic substances.

Problem 1

A number of genetic conditions which are believed to enhance susceptibility to toxic substances are autosomal recessives. While it may be known that the homozygous condition may be at greatly increased risk (e.g. homozygous SAT deficiency, homozygous for AT, XP), little is known about the quantification of risk for the heterozygous state. While the heterozygote will undoubtedly be at markedly lower risk than the homozygous individual, the principal public health question focuses on the heterozygote because of its considerably greater frequency in the population. For example, while the homozygous state for AT occurs with a frequency of about 2.5 in 100 000, the heterozygote occurs with a frequency of about 1 in 100 (Calabrese, 1984). Consequently, a major research challenge is an accurate, rapid, inexpensive and acceptable manner to identify the heterozygote. If this can be accomplished it creates opportunities for a wide variety of epidemiological and clinical studies as well as opportunities to develop in vitro laboratory assays.

Problem 2

There is a general lack of appropriate animal models to predict responses in humans. This results from two principal limitations. First, for an animal model to be an effective qualitative and quantitative predictor of human responses to environmental toxins, it is important to have an excellent understanding of the human condition so that it will be possible to judge whether the animal model appropriately simulates the human condition. Consequently, the search for appropriate predictive animal models involves a concurrent search for improved understanding of the actual human condition one desires an animal model for. This may sound initially contradictory but it isn't. The answer lies in understanding the role of the animal model in the research strategy. The development of an animal model in essence provides a predictive tool in which investigators can evaluate responses to a variety of relevant exposure scenarios that allow various hypotheses to be evaluated. Data from such studies permit biologically based hypotheses to be developed that could be validated via epidemiological methods. Thus, the development of the model should be viewed as an essential tool that leads to improved and more cost-effective epidemiological studies.

The second limitation relates to the current state of biomedical research in general. In the cases of sickle cell trait/anaemia and thalassemia, there is no animal model that adequately represents the human condition. This also appears to be the case for SAT deficiency and DNA repair disease/carrier conditions. For an extensive evaluation of animal models for high risk groups, see Calabrese (1983). While there is little question that progress has been continuously made over the past decades of biomedical research, progress is often painfully slow. Several research strategies present opportunities for progress.

(a) The occurrence of genetic heterogeneity in the human population is mirrored in the animal population in terms of diversity amongst species as well as in adaptations to different environments. The highly variable animal model population needs to be better appreciated since such diversity may be useful as a tool to study the occurrence of human conditions. For example, the range of human enzyme activities for catalase, sulfate conjugation, glucuronidation and others is very broad with certain subgroups of the population being very low in these activities because of genetic factors. It should be appreciated that the animal model kingdom possesses species that reflect the entire spectrum of human activities.

(b) Models have been developed over the decades by irradiation of reproductive cells to evaluate offspring. While successes have been obtained, this has not been a highly fruitful process. It is very likely that genetic engineering practices will become a predominating factor in the development of new animal models (see section 2.6). For example, it would be of considerable ecogenetic research value if human genes for sickle cell trait could be transferred into an experimental model.

Problem 3

The multicausal nature of many complex multistage diseases presents numerous theoretical and practical research concerns (Schulte, 1986). On the theoretical side, investigators often do not know the principal contributing underlying predispositions for the development of environmentally-induced disease, nor the extent of their contributions. This often results in such factors being ignored and only one or two factors being specifically considered. This has often led to contradictory findings and premature rejections of hypotheses based on inadequate understanding. This has been seen in studies of SAT heterozygotes and the occurrence of chronic obstructive pulmonary disease (COPD), acetylator phenotype and bladder cancer association amongst others (Calabrese, 1984).

On the practical side, the recognition and quantification of multicausal factors affecting susceptibility to environmental agents increases the degree of complexity in the conduct of epidemiological studies, often resulting in the need for greater sample sizes, more sophisticated biological monitoring and considerably greater expense. However, such studies would also inherently recognize the need to link epidemiologists with biomedical researchers in the joint conduct of such studies with consequent improvement in the overall research product.

2.9 GENETIC SCREENING: CURRENT STATUS

Calabrese (1984) offered the following conclusion concerning the biological foundations justifying genetic screening procedures and the future role of such procedures. Although the years since have yielded significant advances in limited areas the basic conclusion remains unchanged and is repeated here.

> There is insufficient scientific justification to conduct genetic screening procedures for the purpose of job denial or transfer in order to avoid potential occupationally related diseases. Despite the fact that various industrial practices either in the United States or abroad have at times reported and/or recommended such actions with respect to G–6-PD deficiency, sickle-cell trait, thalassemia, and SAT heterozygotes, there is insufficient evidence at present to justify such programs, policies, and recommendations.
>
> The identification of genetic factors that may predispose the occurrence of job-related disease is a science truly in its infancy. The biological foundations of the concept of genetic screening to identify predispositions to occupational disease are on sound biological grounds. While the future suggests a very widespread and influential role in regulatory processes, the present state of the art can only recommend more study and not the implementation of any job screening programs except as related to the acquisition of data in properly designed studies. Nevertheless, it appears that genetic differences contribute in important ways in explaining the variability of responses to environmental agents. What percentage of the total variability in disease incidence may be explained by genetic factors is uncertain. It should be recognized that other biological variables such as age, nutritional status, preexisting disease, and lifestyle also contribute in highly important ways in affecting the body's susceptibility to a variety of environmental insults. The study of factors affecting susceptibility to environmental and/or occupational diseases therefore should not stop with a

quantification of genetic influences, as important as they may be, but should be comprehensive and incorporate the abovementioned host variables.

2.10 CONCLUSION

There are excellent theoretical foundations to support the premise that genetic factors are likely to enhance the risk of affected individuals reacting adversely to particular environmental agents. However, the degree to which supporting data are available from studies with appropriate predictive animal models and/or human epidemiological studies is highly variable. Recognition of the role of genetic factors in the assessment of environmentally-induced disease helps in the elucidation of the complex web of multifactorial diseases. Acquisition of such information is likely to be an invaluable tool to public health and medical officials as they develop research, treatment and educational programs. In addition, such information is likely to be employed in the risk assessment process used to derive exposure standards for toxic agents found in the workplace and community. However, in order for this field of ecogenetics to realize these projected contributions, a long-range research program needs to be developed and implemented.

REFERENCES

Aderele, W., Ojo, C., Osanyintuji, V. O. and Oduwole, O. (1985) Serum alpha$_1$-antitrypsin levels in asthmatic children. *Afri. J. Med. Sci.*, **14**, 161–7.

Ames, M. M., Selassie, C. D., Woodson, L. C. *et al.* (1986) Thiopurine methyltransferase: structure-activity relationships for benzoic acid inhibitors and thiophenol substrates. *J. Med. Chem.*, **29**, 354–8.

Amoruso, M. A. (1985) Estimation of risk of glucose–6-phosphate dehydrogenase deficient red cells to ozone and nitrogen dioxide. Health Effects Institute Research Report, *Report of the institute's Health Review Committee*, Cambridge, MA.

Anderson, R. J., Weinshilboum, R. M., Phillips, S. F. and Broughton, D. O. (1981) Human platelet phenol sulfotransferase: assay procedure, substrate and tissue correlations. *Clin. Chem. Acta.*, **110**, 157–67.

Ayesh, R., Idle, J. R., Ritchie, J. C. *et al.* (1984) Metabolic oxidation phenotypes as markers for susceptibility to lung cancer. *Nature*, **312**, 169–70.

Beckman, G., Stjernberg, N. L. and Eklund, A. (1984) Is the PiF allele of alpha$_1$-antitrypsin associated with pulmonary disease? *Clin. Genet.*, **25**, 491–5.

Bell, G., Selby, J. and Rutter, W. (1982) The highly polymorphic region near the insulin gene is composed of single tandemly repeated sequences. *Nature*, **295**, 31–5.

Blackwood, R. A., Moret, J., Mandel, I. and Turino, G. M. (1984) Emphysema induced by intravenously administered endotoxin in an alpha$_1$-antitrypsin-deficient rat model. *Am. Rev. Respir. Dis.*, **130**, 231–6.

Bloom, K. E., Brewer, G. J., Magon, A. M. and Wetterstroem, N. (1983) Microsomal incubation test of potentially hemolytic drugs for glucose–6-phosphate dehydrogenase deficiency. *Clin. Pharmacol. Ther.*, **33**(4), 403–9.

Brewer, G. J. (1971) Annotations: human ecology, an expanding role for the human geneticist. *Am. J. Hum. Genet.*, **23**, 92–4.

Brieger, H (1963) Genetic bases of susceptibility and resistance to toxic agents. *J. Occup. Med.*, **5**(11), 511–15.

Calabrese, E. J. (1983) *Principles of Animal Extrapolation*, Wiley Interscience, New York.

Calabrese, E. J. (1984) *Ecogenetics: Genetic Variation in Susceptibility to Environmental Agents*, Wiley Interscience, New York.

Calabrese, E. J. (1986) Ecogenetics: historical foundation and current status. *J. Occup. Med.*, **28**(10), 1096–1102.

Calabrese, E. J. (1991) *Predicting Toxicological Hazards of Multiple Chemical Interactions.* Lewis, London.

Calabrese, E. J., Horton, H. M. and Leonard, D. A. (1987a) The effects of four steroids on glucose–6-phosphate dehydrogenase activity of human and C57L/J mouse erythrocytes. *J. Environ. Sci. Health*, **A22**(6), 563–74.

Calabrese, E. J., Horton, H. M. and Leonard, D. A. (1987b) The effect of the steroid dehydroepiandrosterone on the glucose–6-phosphate dehydrogenase activity of human erythrocytes. *J. Environ, Sci. Health*, **A21**(6), 499–511.

Campbell, N. R. C., Dunnette, J. H., Mwaluko, G. *et al.* (1984) Platelet phenol sulfotransferase and erythrocyte catechol-O-methyltransferase activities. Correlation with methyldopa metabolism. *Clin. Pharmacol. Ther.*, **35**, 55–63.

Carta, P., Anni, M. S., Glacomina, C. *et al.* (1987) Occupational lead exposure, G–6-PD deficiency and ß-thalassemic trait. *Med. Lav.*, **78**, 75–85.

Cartwright, R. A., Glashan, R. W., Rogers, H. J. *et al.* (1982) Role of *N*-acetyltransferase phenotypes in bladder carcinogenesis. A pharmacogenetic epidemiological approach to bladder cancer. *Lancet*, **ii**, 842–5.

Castro, O., Orlin, J., Rosen, M. W. and Finch, S. C. (1973) Survival of human sickle cell erythrocytes in heterologous species: response to variations in oxygen tension. *Proc. Natl Acad. Sci. USA*, **70**, 2356–9.

Christensson, P., Arborelius, M. J. and Kautto-Wiberg, R. (1986) Changes of volume of trapped gas after bronchodilation in subjects with suspected subclinical emphysema. *Respiration*, **49**, 16–22.

Conney, A. H. (1982) Induction of microsomal enzymes by foreign chemicals and carcinogenesis by polycyclic aromatic hydrocarbons: G. H. A. Clowes Memorial Lecture. *Cancer Res.*, **42**, 4875–4917.

de Bruin, A. (1976) *Biochemical Toxicology*, Elsevier, Amsterdam.

De Weck, A. L. (1983) Immunopathological mechanisms and clinical aspects of allergic reactions to drugs, in *Allergic Reactions to Drugs* (eds. A. L. De Weck and H. Bundgaard), Springer-Verlag, Berlin, pp. 75–133.

Eales, L. J. Parkin, J. M., Forster, S. M. *et al.* (1987) Association of different allelic forms of group specific component with susceptibility to and clinical manifestation of human immunodeficiency virus infection. *The Lancet*, **i**, 999–1002.

Eriksson, S., Lindell, S. E. and Wiberg, R. (1985) Effects of smoking and intermediate alpha$_1$-antitrypsin deficiency (P$_1$MZ) on lung function. *Eur. J. Resp. Dis.*, **67**, 279–285.

Gladwell, M. (1988) DuPont to sell genetically engineered mice that carry cancer genes. *The Boston Globe*, 16 November, p. 4.

Goldstein, R. A. and Patterson, R. (1984) Summary. *J. Allergy Clin. Immunol.*, **74**, 643–4.

Gonzalez, F. J., Jaiswal, A. K. and Nebert, D. W. (1987) P–450 genes: evolution, regulation, and relationship to human cancer and pharmacogenetics. *Cold Spring Harbor Symp. Quant. Biol.*, **51**, 879–90.

Gonzalez, F. J., Skoda, R. C., Kimura, S. *et al.* (1988) Characterization of the common genetic defect in humans deficient in debrisoquine metabolism. *Nature*, **331**, 442–6.

Grammer, L. C., Roberts, M., Nicholls, A. J. *et al.* (1984) IgE against ethylene oxide-altered human serum albumin in patients who have had acute dialysis reactions. *J. Allergy Clin. Immunol.*, **74**, 544–6.

Horn, S. L., Tennet, R. K., Cockcroft, D. W. *et al.* (1986) Pulmonary function in PiM and NZ grain workers. *Chest*, **89**, 795–9.

Horton, H. M. and Calabrese, E. J. (1986a) Predictive models for human G–6-PD deficiency. *Drug Metab. Rev.*, **17**, 261–81.

Horton, H. M. and Calabrese, E. J. (1986b) Attempts to validate a possible predictive animal model for human erythrocyte G–6-PD deficiency. *J. Environ. Sci. Health*, **A21**(3), 235–47.

Hultin, T. A. (1983) Metabolism and toxicity in the genetic mouse model, doctoral dissertation, University of Michigan, Ann Arbor.

Hultin, T. A. and Weber, W. W. (1986) Genetic differences in inhibition of 2-aminofluorene *N*-acetyltransferase activity between C57BL/6J and A/J mice. *Biochem. Pharmacol.*, **35**, 1214–16.

Idle, J. (1988) Enigmatic variations. *Nature*, **331**, 391–2.

Islam, S. I., Idle, J. R. and Smith, R. L. (1980) The polymorphic 4-hydroxylation of debrisoquine in a Saudi Arab population. *Xenobiotica*, **10**, 819–26.

Jeffreys, A., Wilson, V. and Thein, S. (1985a) Hypervariable 'minisatellite' regions in human DNA. *Nature*, **314**, 67–73.

Jeffreys, A., Wilson, A. and Thein, S. (1985b) Individual-specific 'fingerprints' of human DNA. *Nature*, **316**, 76–9.

Jensen, W. N. (1962) Hereditary and chemically induced anemia. *Arch. Environ. Health*, **5**, 212–16.

Kalow, W. (1962) *Pharmacogenetics: Heredity and the Response to Drugs*, Saunders, Philadelphia.

Kamataki, R., Maeda, K., Yamazoe, Y. *et al.* (1983) A high-spin form of cytochrome P–450 highly purified from polychlorinated biphenyl-treated rats. *Mol. Pharmacol.*, **24**, 146–55.

Karol, M. H., Loset, H. H. and Alarie, Y. C. (1978) Totyl-specific IgE antibodies in workers with hypersensitivity to toluene diisocyanate. *Am. Ind. Hyg. Assoc. J.*, **39**, 454–8.

Keith, R. A., Van Loon, J., Wussow, L. F. and Weinshilboum, R. M. (1983) Thiol methylation pharmacogenetics: heritability of human erythrocyte thiol methyltransferase activity. *Clin. Pharmacol. Ther.*, **34**, 521–8.

Kerremans, A. L., Lipshy, J. J., Van Loon, J. *et al.* (1985). Cephalosporin-induced hypopropthrombinemia: possible role for thiol methylation of 1-methyltetrazole–5-thiol and 2-methyl–1,3,4-thiadiazole–5-thiol. *J. Pharmacol. Exp. Ther.*, **235**, 382–8.

Khoury, M. J., Beaty, T. H., Tockman, M. S. *et al.* (1985). Familial aggregation in chronic obstructive pulmonary disease: use of the loglinear model to analyse intermediate environmental and genetic risk factors. *Genet. Epidemiol.*, **2**, 155–66.

Kouri, R. E., Lubet, R. A., McKinney. C. E. *et al.* (1983) Aryl hydrocarbon hydroxylase activity 'of mice and humans', in *Applications of Biological Markers to Carcinogen Testing* (eds H. Melman and S. Sell), Plenum Publishing, New York, pp. 179–97.

Kouri, R. E., McLemore, T., Jaiswal, A. K. and Nebert, D. W. (1986) Current cellular assays for measuring clinical drug metabolizing capacity-impact of new molecular biologic techniques, in *Ethnic Differences in Reaction to Drugs and Xenobiotics*, Alan R. Liss, New York, pp. 453–69.

Kupper, A., Schmid, B. and Pfaff, G. (1986) Pharmacogenetics of dextromethorphan o-demethylation in man. *Xenobiotica*, **16**(5), 421–33.

Lappe, M. (1988) Ethical issues in genetic screening for susceptibility to chronic lung disease. *J. Occup. Med.*, **30**, 493–501.

Lennard, L., Van Loon, J., Lilleyman, J. *et al.* (1986) Thiopurine pharmacogenetics: correlation of thiopurine methyltransferase (TPMT) activity and 6-thioguanine nucleotide (6-TGN) levels. *Clin. Pharmacol. Ther.*, **39**, 206.

Lennard, L., Van Loon, J. A., Lilleyman, J. S. and Weinshilboum, R. M. (1987) Thiopurine pharmacogenetics in leukemia: correlation of erythrocyte thiopurine methyltransferase activity and 6-thioguanine nucleotide concentrations. *Clin. Pharmacol. Ther.*, **41**, 18–25.

Marx, J. L. (1988) Gene transfer is coming on target. *Science*, **242**, 191–2.

Mason, T. J., Prorok, P. C. and Sostlow, R. D. (eds) (1986) Conference on medical screening and biological monitoring for the effects of exposure in the workplace. *J. Occup. Med.*, **28**(10), 913–1126.

Mathews, K. P. (1984) Clinical spectrum of allergic and pseudoallergic drug reactions. *J. Allergy Clin. Immunol.*, **74**, 558–66.

Mattano, S. S. and Weber, W. W. (1987) Kinetics of arylamine *N*-acetyltransferase in tissue from rapid and slow acetylator mice. *Carcinogenesis*, **8**, 133–7.

Meier, H. (1963) *Experimental Pharmacogenetics*, Academic Press, New York.

Menkes, H. A., Cohen, B. H., Beaty, T. H. *et al.* (1984) Risk factors, pulmonary function, and mortality. *Prog. Clin. Biol. Res.*, **147**, 501–21.

Motulsky, A. (1957) Drug reactions, enzymes, and biochemical genetics. *JAMA*, **165**, 835–7.

Mountain, J. T. (1963) Detecting hypersusceptibility to toxic substances. *Arch. Environ. Health*, **6**, 357.

Nebert, D. W., Adesnik, M., Coon, M. J. *et al.* (1987) The P–450 gene superfamily: recommended nomenclature. *DNA*, **6**, 1–11.

Nebert, D. W., Eisen, H. J. Negishi, M. *et al.* (1981) Genetic mechanisms controlling the induction of polysubstrate monooxygenase (P–450) activities. *Ann. Rev. Pharmacol. Toxicol.*, **21**, 431–62.

Nebert, D. W. and Gonzales, F. J. (1987) P–450 genes: structure, evaluation, and regulation. *Ann. Rev. Biochem.*, **56**, 945–93.

Nebert, D. W., Negishi, M., Lang, M. A. *et al.*(1982) The Ah locus, a multigene family necessary for survival in a chemically adverse environment: comparison with the immune system. *Adv. Genet.*, **21**, 1–52.

Omenn, G. (1986) Susceptibility to occupational and environmental exposures to chemicals, in *Ethnic Differences in Reactions to Drugs and Xenobiotics*, Alan R. Liss, New York, pp. 527–45.

Park, B. K., Coleman, J. W. and Kitteringham, N. R. (1987) Drug disposition and drug hypersensitivity. *Biochem. Pharmacol.*, **36**, 581–90.

Parker, C. W. (1982) Allergic reactions in man. *Pharmacol. Rev.*, **34**, 85–104.

Patterson, R. and Anderson, J. (1982) Allergic reactions to drugs and biologic agents. *JAMA*, **248**, 2637–45.

Patterson, R., Pateras, V., Grammer, L. C. and Harris, K. E. (1986) Human antibodies against formaldehyde–human serum albumin conjugates or human serum albumin in individuals exposed to formaldehyde. *Int. Arch. Allergy Appl. Immunol.*, **79**, 53–9.

Pazmino, P. A. and Weinshilboum, R. M. (1978) Human erythrocyte phenol-*O*-

methyltransferase: radiochemical microassay and biochemical properties. *Clin. Chim. Acta*, **89**, 317–29.

Pelkonen, O. and Nebert, D. W. (1982) Metabolism of polycylic aromatic hydrocarbons: etiologic role in carcinogenesis. *Pharmacol. Rev.*, **34**, 189–222.

Pohl, L. R., Satoh, H., Christ, D. D. and Kenna, J. G. (1988) The immunologic and metabolic basis of drug hypersensitivities. *Ann. Rev. Pharm.*, **28**, 367–87.

Price-Evans, D. A. (1963) Pharmacogenetics. *Am. J. Med.*, **34** 639–62.

Raymond. F. A. and Weinshilboum, R. M. (1975) Microassay of human erythrocyte catechol-*O*-methyltransferase: removal of inhibitory calcium with chelating resin. *Clin. Chim. Acta*, **58**, 185–94.

Reilly, D. K., Rivera-Calimlin, L. and Van Dyke, D. (1980) Catechol-*O*-methyltransferase activity: a determinant of levodopa response. *Clin. Pharmacol. Ther.*, **28**, 278–86.

Reiter, C. and Weinshilboum, R. M. (1982) Platelet phenol sulfotransferase activity: correlation with sulfate conjugation of acetaminophen. *Clin. Pharmacol. Ther.*, **32**, 612–21.

Schneider, C. H. (1983) Immunochemical basis of allergic reactions to drugs, in *Allergic Reactions to Drugs* (eds A. L. De Weck and H. Bundgaard), Springer-Verlag, Berlin, pp. 3–36.

Schulte, P. A. (1986) Simultaneous assessment of genetic and occupational risk factors. Presented at the Fifth International Symposium, Epidemiology in Occupational Health, Los Angeles, 11 September 1986.

Sladek-Chelgren, S. and Weinshilboum, R. M. (1981) Catechol-*O*-methyltransferase biochemical genetics: human lymphocyte enzyme. *Biochem. Genet.*, **19**, 1037–53.

Spielberg, S. P. (1984) In vitro assessment of pharmacogenetic susceptibility to toxic drug metabolites in humans. *Fed. Proc.*, **43**, 2308–313.

Stanbury, J. B., Wyngaarden, J. B. and Fredrickson, D. S. (eds) (1983) *The Metabolic Basis of Inherited Disease*, McGraw-Hill, New York.

Stokinger, H. E. (1962) New concepts and future trends in toxicology. *Am. Ind. Hyg. Assoc. J.*, **23**, 8.

Stokinger, H. E. and Mountain, J. T. (1963) Test for hypersusceptibility to hemolytic chemicals. *Arch. Environ. Health*, **6**, 495–502.

Stokinger, H. E. and Scheel, L. D. (1973) Hypersusceptibility and genetic problems in occupational medicine: A consensus report. *J. Occup. Med.*, **15**, 564–73.

Stokinger, H. E., Mountain, J. T. and Scheel, L. D. (1968) Pharmacogenetics in the detection of the hypersusceptible worker. *Ann. N.Y. Acad. Sci.*, **151**, 968–76.

Vogel, F. (1959). Moderne probleme de humangenetik. (Modern problems of human genetics.) *Ergeb. Inn. Med. Kinderheilk.*, **12**, 52–125.

Waterman, M. R. and Estabrook, R. W. (1983) Induction of microsomal electron transport. *Mol. Cell. Biochem.*, **53/54**, 267–78.

Weber, W. W. (1984) Acetylation pharmacogenetics experimental models for human toxicity. *Fed. Proc.*, **43**, 2332–7.

Weber, W. W. (1987) *The Acetylator Genes and Drug Response*, Oxford University Press, New York.

Weber, W. W., Hultin, T. A. and Morgott, D. M. (1984) Acetylation pharmacogenetics: profiles of arylamines pharmacology and toxicology in a genetic mouse model. *Biochem. Soc. Trans.*, **12**, 87–90.

Weinshilboum, R. M. (1984) Human pharmacogenetics of methyl conjugation. *Fed. Proc.*, **43**, 2303–7.

Weinshilboum, R. M. (1987) Pharmacogenetics of methyl conjugation and thiopurine drug toxicity. *Bioessays*, **7**(2), 78–82.

Weinshilboum, R. M. and Raymond, F. A. (1977) Inheritance of low erythrocyte catechol-*O*-methyltransferase activity in man. *Am. J. Hum. Genet.*, **29**, 125–35.

Weinshilboum, R. M., Raymond, F. A. and Pazmino, P. A. (1978) Human erythrocyte thiopurine methyltransferase: radiochemical microassay and biochemical properties. *Clin. Chim. Acta*, **85**, 323–33.

Weinshilboum, R. M. and Sladek, S. L. (1980) Mercaptopurine pharmacogenetics: monogenic inheritance of erythrocyte thiopurine methyltransferase activity. *Am. J. Hum. Genet.*, **32**, 651–62.

White, R. (1985) DNA sequence polymorphisms revitalize linkage approaches in human genetics. *Trends Genet*, **1**, 177–81.

Wolf, C. R. (1986) Cytochrome P–450s: polymorphic multigene families involved in carcinogen activation. *Trends Genet.*, **2**, 209–14.

Woodson, L. C., Ames, M. M., Selassie, C. D. *et al.* (1983) Thiopurine methyltransferase: aromatic thiol substrates and inhibition by benzoic acid derivative. *Mol. Pharmacol.*, **24**, 471–8.

Zavon, M. R. (1962) Modern concepts of diagnosis and treatment in occupational medicine. *Am. Ind. Hyg. Assoc. J.*, **23**, 30.

3 Genetic predisposition to environmental toxic agents: detection by distribution of ecogenetic markers

N. P. Bochkov and N. V. Titenko

National Research Centre of Medical Genetics, Academy of Medical Sciences, Moscow, USSR

3.1 SUMMARY

Due to past environmental conditions, social structure and/or isolation, human populations exhibit considerable genetic variation in susceptibility to industrial chemicals. With more than 100 ethnic groups in the USSR, these variations are extensive. Glucose–6-phosphate dehydrogenase, cholinesterase and alpha$_1$-antitrypsin show major differences in gene frequencies, thus resulting in different prevalences of predisposition to effects of certain chemicals. In addition, genetic markers in patients with asbestosis, silicosis and other occupational diseases show important, though yet poorly understood, differences from the background rates. Thus, genetic factors are of major significance for individual predisposition to toxic effects of chemicals. Elucidation of the mechanisms involved and the social and legal aspects is an important task for the future.

3.2 INTRODUCTION

Various environmental effects have been the leading factors of selection throughout the development of human populations. For each individual, contacts with the environment take place through numerous chemical reactions in the metabolism of food products and drugs and in the detoxication of exogenous poisons and environmental pollutants. The differences in enzyme activity determine the peculiarities of metabolism for all substances in the body. This means that for different biochemical systems, two or more gene variants may determine the spectrum of isoenzymes or protein variants that normally do not necessarily have any pathological effect and may to a certain extent influence and/or contribute to normal variations in metabolic reactions. Remarkable differences in common human population polymorphism is caused by geographic location, ethnic origin and an adaptation to the various environmental factors. Rare variants in populations (less than 1%) also occur, caused by either new mutations or segregation process, which can be not only physiologically neutral but result in some pathology.

Practically every human population affected by any factor can be divided into stable, conditionally stable and sensitive individuals, depending on the individual reaction to the exposure. The distribution of individuals that react can be normal, bimodal or more complex.

The individual can be sensitive to one factor and non-sensitive to another, depending on the inherited alleles. The frequencies of concrete ecogenetically expressed alleles in different populations can be diverse. High-concentration pathological alleles in selected populations might be explained by specific environmental conditions in the past, social structure and/or isolation. The sensitivity of particular individuals to environmental factors can manifest itself in different ways, such as acute toxic reactions (poisoning), specific occupational pathology, allergy, immunodeficiency, teratogenicity, cancerogenicity and mutagenicity. Without doubt, the significance of genetically determined biochemical differences in individuals is a cause of sensitivity to various environmental agents. Table 3.1 shows examples of genetic predisposition to pathological effects of environmental factors. Finding the true distribution of ecogenetic markers in populations of the USSR is a very important task. The USSR is characterized by variable climatic conditions and includes more than 100 ethnic groups. Geographic and ethnic variabilities are maximal compared with other regions or countries in the world.

3.3 GLUCOSE–6-PHOSPHATE DEHYDROGENASE

Glucose–6-phosphate dehydrogenase (G–6-PD) deficiency, created by epidemics of tropical malaria, is a classic example of ecogenetic polymorphism. Mutant alleles of G–6-PD (Gd^-) are well known to have a modifying effect on reactions not only to anti-malarial drugs and sulfonamides but also to hypoxia and different chemicals. Table 3.2 shows the G–6-PD deficiency frequency in various populations and ethnic groups of the USSR (Krasnopolskaya and Shatskaya, 1987). The maximal G–6-PD

Table 3.1 Examples of genetic predisposition to pathological conditions influenced by environmental factors

Type of effect	*Pathological condition*	*Provoking factor*	*Reference*
Acute poisoning	Severe hyperthermia	Inhalation anaesthetics	Goedde, 1986
	Haemolysis	Vicia fava, antimalarial drugs, lead	
	Breath arrest	Some drugs	
Allergy	Dermatitis	Beryllium	Burnazjan, 1968
	Bronchial asthma	Chlorinated hydrocarbons	Davies and Blainey, 1984
Carcinogenesis	Mesothelioma	Asbestos	McDonald and MacDonald, 1981
	Lung cancer		
	Bladder cancer	Benzidine	Brown, 1986
Occupational diseases	Asbestosis	Asbestos	WHO, 1986
	Silicosis	Coal dust	de Shazo, 1982
Mutagenesis	Elevated level of chromosomal aberrations	UV light	Cohen and Levy, 1989
		Ionizing radiation	
		Chemical mutagens	
Teratogenesis	Congenital malformations	Thalidomide	Lenz, 1985

Table 3.2 The frequency of G–6-PD deficiency in different populations and ethnic groups in the USSR (Krasnopolskaya and Shatskaya, 1987)

Group	*No. of individuals investigated*	*Frequency of* Gd^{-}
Azerbaijanians (Sheky)	1488	0.10
Azerbaijanians (Koby)	469	0.04
Turkmens	869	0.01
Tajiks	1206	0.01
Ashkenazi Jews (Gomel)	515	0.009
Uzbeks	1652	0.005
Russians (Kostroma)	952	0.004

deficiency frequency (corresponding to the world maximum) was found in Azerbaijan, where a wide spectrum of different alleles is present. New alleles have also been found in Russian populations. In that investigation 69 individuals with mutant G–6-PD forms were studied; 27 different variants were found, 20 of which had not been described before and 3 of which were only partially characterized according to the WHO programme (WHO, 1967). Results show that the alleles found among Ashkenazi Jews and in various Azerbaijan subpopulations can be grouped into related groups or clusters. The detection of the allele families with identical phenotypical expression among Azerbaijanian, Ashkenazi Jews and other populations living on a permanent basis in malaria-stricken areas makes it possible to speculate about the similarity of their genetic origin.

3.4 SERUM CHOLINESTERASE

Polymorphism of human serum cholinesterase, determined by two autosomal loci, E_1 and E_2, was examined among healthy Russians (Mheidze,1986) and Azerbaijans (Hajiyev and Sergeyev, 1987) and also in groups of patients with schizophrenia, peptic ulcer, hereditary erythrocytopathies, tuberculosis, thyrotoxicosis, essential hypertension and rheumatic disease (Hajiyev *et al.*, 1987). Table 3.3 summarizes these data. The differences in enzyme variant frequencies were not significant for children (165), donors (750) and persons aged 60 years and older (120) in the Russian population of Leningrad. The frequency of E_1 variants was significantly higher among Azerbaijanian pupils, as compared to the donors and children of Leningrad. This was conditioned by high frequencies of mutant alleles of the E_1^a, E_1^f and E_1^s loci responsible for atypical and 'silent' enzyme variants. The increased frequencies of E_1^a and E_1^f alleles in the patients with peptic ulcer (12.3%), hereditary erythrocytopathies (23.2%), schizophrenia (2.8%) and tuberculosis (5.4%), as compared to the control group, were statistically significant.

3.5 ALPHA$_1$-ANTITRYPSIN

Another example of the ecogenetic marker is the proteinase-alpha–1-inhibitor (PI), which is concerned with a pathogenesis of bronchopulmonary and liver diseases. The ecogenetic significance of S and Z variants of PI in the development of obstructive lung disease was demonstrated in the early 1960s. Since then the spectrum of allele variants has expanded to over 60. Recent data indicate a wider effect of PI on other pathologies (Cox, 1986; Michalski *et al.*, 1986; Basys, 1987).

Table 3.3 Serum cholinesterase polymorphism among Russians and Azerbaijanians

		Frequencies of E_1 alleles				
Group	No. of individuals observed	E_1^u	E_1^d	E_1^f	E_1^s	Reference
Russians (control)	1035	0.980	0.012	0.008	–	Mheidze, 1986
Azerbaijanians (control)	933	0.968	0.022	0.004	0.006	Hajiyev and Sergeev, 1987
Azerbaijanians with:						
schizophrenia	213	0.953	0.021	0.023	0.02	
erythrocytopathies	136	0.930	0.066	0.004	–	
peptic ulcer	220	0.923	0.061	0.011	0.005	Hajiyev *et al.*, 1987
tuberculosis	240	0.956	0.010	0.027	0.006	
thyrotoxicosis	135	0.993	0.004	0.004	–	
hypertension	168	0.985	0.009	–	0.006	
rheumatic diseases	96	0.990	0.010	–	–	

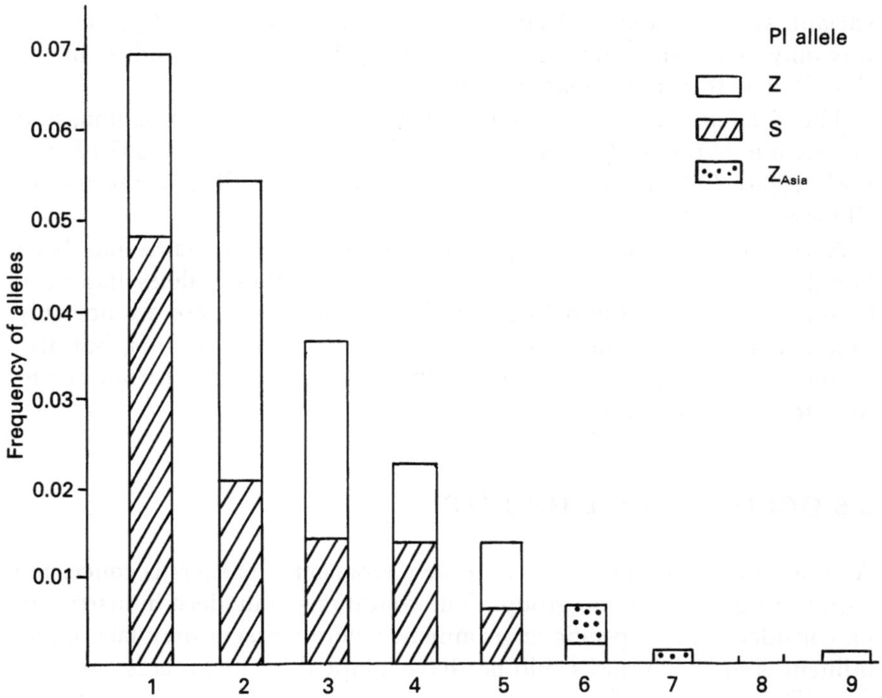

Figure 3.1 Geographic distribution of pathological PI alleles from northwest to southeast Eurasia. The numbers 1 to 9 show the frequencies of S and Z alleles in different populations: 1. English (Arnaud *et al.*, 1979); 2. Swedish (Frants and Eriksson, 1980); 3. Estonians (Titenko and Parik, 1989); 4. Russians (Titenko and Parik, 1989); 5. Ukrainians (Chernenko *et al.*, 1987 personal communication); 6. Buryats (West to Baikal Lake) (this chapter); 7. Buryats (East to Baikal Lake) (this chapter); 8. Mongolians (Ziujin *et al.*, 1986); 9. Chinese (Ziujin *et al.*, 1986).

We collected data on the frequencies of pathological S and Z variants in various populations of the USSR and other countries (Fig. 3.1). Blood samples (more than 3500) were drawn from Russians, Estonians, Ukranians and Buryats. The incidence of pathological PI variants decreased in the following order: Estonians, Russians, Ukrainians, West Baikal Buryats and East Baikal Buryats. The difference in the incidence of pathological PI variants between these populations reached a factor of eight. Among more than 900 Buryats studied in Baikal Lake region only a few Z variants were found and the S variant was absent. Moreover, in the West Baikal population, Z variant, which is common in European populations, existed only in trace frequency (0.0017). A new variant, slower in the isoelectrofocusing picture, called the Z_{Asia} variant, was only established among the more than 500 samples studied in the East Baikal Buryat population (0.0027).

The Z allele has not been found in three populations in China, and in only one of them (Zhuang) the gene frequency of S was 0.007 (Ziujin *et al.*, 1986). Qilong and Liang (1983) found neither PI Z nor S gene in Chinese populations.

A very low population frequency of pathological variants has been found in China while the highest frequency of these alleles has been found in Scandinavia and England. This means that there are not only differences in the frequencies and the spectrum of PI variants, but also a tendency to increase according to the decreased gradient from northwest to southeast Eurasia.

3.6 OCCUPATIONAL GROUPS

Another approach in the search for new ecogenetic markers is concerned with testing occupational groups. The conditions of modern industry can be considered as a specific environment where workers may face many influences to which they could not have adapted in the process of evolution. The spectrum of ecogenetic reactions in occupational contacts can be fairly wide, and interest in this area of human genetics is growing.

The distribution of ecogenetic markers among workers in contact with asbestos has been studied. Asbestos can cause asbestosis in 3–12% of workers (WHO, 1986). All 621 workers tested were Russians from the Moscow region. We compared five groups who differed on level and time of exposure:

1. patients wth asbestosis;
2. persons with absence of any pattern of occupational disease caused by asbestos after long exposure (10–38) years;
3. persons working at the plant less than 10 years;
4. workers with time of middle-level exposure less than 1 year; and
5. control-group residents of the city.

We considered the people of the first group as susceptible to asbestos and the second group as comparatively non-susceptible to this factor. In the framework of a combined programme, more than 20 polymorphisms (mainly biochemical) were investigated.

Significant differences in the distribution of markers between the asbestos-exposed group (group 1) and non-asbestos-exposed group (group 2) were established in at least three genetic markers (Table 3.4). Among the patients, Hp1, Rh^- and rare PI variants (together with M3 and S) were significantly increased. The data on the pathological effect of the M3 allele of the PI gene are in line with other results showing the possible significance of the M3 allele in the pathogenesis of lung tuberculosis, bladder cancer and chronic non-specific lung disorders in children (Basys, 1987; Chernenko *et al.* 1987, personal communication).

Table 3.4 Distribution of some ecogenetic markers in the asbestos-exposed group (1) and non-asbestos-exposed group (2) in a comparison of five groups differing on level and time of exposure

Genetic polymorphism	*Group*	*No. of individuals*	*Frequency of gene*	*Significance*
Hpl	1	44	0.534	2.32 ($p<0.05$)
	2	68	0.316	
d(Rh$^-$)	1	44	0.403	
	2	68	0.174	2.60 ($p<0.01$)
PI	1	38	0.171	
(S + Var + M3)	2	68	0	2.36 ($p<0.05$)

A similar pattern has been observed in the haptoglobin Hpl allele. The frequency of Hpl is significantly higher among people with various chronic diseases (Frölander, 1988).

Hereditary predisposition to silicosis is not completely clarified. The possible associations between different antigens in the HLA system and silicosis were analyzed among Russian miners exposed to coal and other mineral dust (Nazarova *et al.*, 1980). The distribution of 11 antigens of A and B groups of the HLA system was studied in 87 patients with silicosis in stages I, II and III (time of exposure 10–25 years) and in a control group ($n = 540$). The B antigen (HLA-B13) was significantly associated with silicosis. The frequency of the gene in patients was 0.097 and in controls 0.024. The relative risk of acquiring the disease in association with HLA-B13 was 4.5.

The study of the role of ecogenetic markers in occupational pathology is just beginning. Of 120 workers affected by photoactive agents in Sweden, 73 developed face eczema (Beckman *et al.*, 1985). This group displayed a sharp increase of allele C2 of the Tf gene, which is presumably more active in forming hydroxyl radicals resulting in cell damage.

A diverse reaction of various phenotype carriers of blood serum alkaline phosphatase has been found in the USSR among workers in contact with organochlorine pesticides (Dolmatova and Rafikov, 1985). Workers with the phenotype Pp2 and who had worked for a long time were found to have an increased activity of alkaline phosphatase, whereas those with phenotype Pp3 showed a decrease, which suggests a genetically determined mechanism of the reaction to occupational pullutants.

3.7 DISCUSSION

Ecogenetic data indicate that relatively few occupational pollutants cause pathological reactions in certain individuals exposed to comparatively low levels. New screening programmes are needed to identify such individuals with potential predisposition to occupational disorders. The creation of such a system in the future will enable an individual to choose a minimal risk job.

This proposal will have to address technical, legal and social problems. Much work lies ahead in searching for new ecogenetic markers for a variety of occupational hazards. The following criteria are recommended for ecogenetic implementation in plants and factories: the number of persons who develop an occupational disease is relatively small compared to the total force; the disease is severe or has irreversible complications; the disease is linked to specific occupational factors.

Future development of human ecogenetics will be mainly connected with advances in the study of biochemical polymorphism in populations and in thorough genetic analysis of various abnormal reactions caused by occupational agents.

REFERENCES

Arnaud, P., Galbraith, R. M., Faulk, W. P. and Black C. (1979) PI phenotypes of alpha$_1$-antitrypsin in southern England: identification of M subtypes and implications for genetic studies. *Clin. Genet.*, **15**, 406–10.

Basys, V. (1987) Concentration and typing of the alpha$_1$-antitrypsins in healthy and sick children, Summary report of the annual meeting of the Study Group for Paediatric Laboratory Diagnostics, Erfurt, 233–37 (in German).

Beckman, L., Beckman, G., Adergren, B. *et al.* (1985) Transferrin C subtypes and occupational photodermatosis of the face. *Hum. Hered.*, **35**, 89–94.

Brown, S. L. (1986) Differential susceptibility: implications for epidemiology, risk assessment, and public policy, in *Phenotypic Variation in Populations. Relevance to Risk Assessment* (eds. A. D. Woodhead, M. A. Bender and R. C. Leonard), Basic Life Sciences, Vol. 1, Plenum Press, New York, pp. 255–69.

Burnazjan, A. I. (ed.) (1968) *Clinics of Chronic and Acute Damages by Berilium*, Medicina, Moscow.

Cohen, M. M. and Levy, H. P.(1989) Chromosome instability syndromes, in *Advances in Human Genetics* (eds H. Harris and K. Hirschhorn), Plenum Press, New York, pp. 43–149.

Cox, D. W. (1986) Clinical and molecular studies of alpha$_1$-antitrypsin deficiency, in *Ethnic Differences in Reactions to Drugs and Xenobiotics*, Alan R. Liss, New York, pp. 373–84.

Davies, R. J. and Blainey, D. A. (1984) Occcupational asthma classification and clinical aspects. *Sem. Resp. Med.*, **5**, 229–39.

de Shazo, R. D. (1982) Current concepts about the pathogenesis of silicosis and asbestosis. *J. Allergy*, **70**, 41–9.

Dolmatova, I. Yu. and Rafikov, Kh. S. (1985) Phenotype modification peculiarities of serum alkaline phosphatase activity in the blood serum of workers handling organochlorine compounds. *Gig. Tr. Prof. Zabol.*, **10**, 18–22.

Frants, R. R. and Eriksson, A. W. (1980) PI subtypes of alpha$_1$-antitrypsin in isolate studies, in *Population Structure and Genetic Disorders* (eds R. R. Frants, A. W. Eriksson, H. R. Forsins *et al.* Academic Press, London, pp. 199–210.

Fröhlander, N. (1988) The haptoglobin polymorphism. Impact of Hp phenotypes on multifactorial disorders, thesis, Umea University.

Goedde, H. W. (1986) Ethnic differences in reactions to drugs and other xenobiotics. Outlook of a geneticist, in *Ethnic Differences in Reactions to Drugs and Xenobiotics* (eds W. Kalow, H. W. Goedde and D. P. Agarwal), Alan R. Liss, New York, pp. 9–20.

Gregor, A, Singh S., Turner-Wasuide, M. *et al.* (1979) The role of histo-compatibility (HLA) antigens in asbestosis. *Br. J. Dis. Chest.*, **73**, 245–52.

Hajiyev, B. O. and Sergeyev, A. S. (1987) Biochemical polymorphism of serum cholinesterase in populations of Kyurdamir district of the Azerbaijan SSR. *Genetika (Rus)*, **23**, 549–58.

Hajiyev, B. O., Sergeyev, A. S., Movsum-Zade, K. M. (1987), Associations between mutant alleles of serum cholinesterase and some multifactorial and infectious diseases. *Genetika (Moskva)*, **23**, 559–63.

Krasnopolskaya, K. D. and Shatskaya, T. L. (1987) Distribution of Gd$^-$ alleles in some ethnic groups of the USSR. *Hum. Genet.*, **75**, 258–63.

Lenz, W. (1985) Thalidomide embryopathy in Germany, 1959–1961, in *Prevention of Physical and Mental Congenital Defects: Basic and Medical Science, Education, and Future Strategies* (ed. M. Marois), Alan R. Liss, New York, pp. 77–83.

McDonald, J. C. and McDonald, A. D. (1981) Mesothelioma as a index of asbestos impact, in *Quantification of Occupational Cancer* (eds R. Peto and M. Schneiderman), Cold Spring Harbor Laboratory, (Banbury Report 9), pp. 73–82.

Mheidze, M. O. (1986) Polymorphism of human serum cholinesterase (CHE, EC 3.1.1.8) among populations of Leningrad and Azerbaijan. *Genetika (Moskva)*, **12**, 2852–9.

Michalski, J. P., McCombs, C. C., Scopelitis, E. *et al.* (1986) Alpha-antitrypsin phenotypes, including M subtypes, in pulmonory disease associated with rheumatoid arthritis and systemic sclerosis. *Arthritis Rheum.*, **29**(5), 586–91.

Nazarova, E. K., Fedrunova, V. S., Pevnitsky, L. A. and Glotova, K. V. (1980) Association of HLA-B13 antigen with silicosis. *Immunology (Rus.)*, **5**, 22–5.

Qilong, Y. and Liang, C. (1983) The distribution of gene frequency of serum alpha$_1$-antitrypsin in Chinese. *Sci Sin.*, **N10**, 897–903.

Titenko, N. V. and Parik J. I. (1989) Gradient of serum alpha-proteinase inhibitor pathological variants in Eurasia, in *Problems of Establishing Human Heredity on the Level of Protein Products of Gene Expression* (ed. S. S. Shyshkin), Medicina, Moscow, pp. 68–70.

WHO (1967) *Standardization of Procedures for the Study of Glucose-6-Phosphate Dehydrogenase*, Technical Report Series, No. 366, World Health Organization, Geneva, pp. 1–53.

WHO (1986), *Asbestos and other Natural Mineral Fibres*, Environmental Health Criteria 53, World Health Organization, Geneva, pp. 108–43.

Ziujin X., Meiying, C., Shizhe, L. *et al.* (1986) Polymorphisms of PI, Hp, ADA and AK in Mongolian, Korean and Zhuang populations of China. *Ann. Hum. Biol.*, **13**(N3), 245–51.

4 Genetic predisposition to common diseases

K. Berg

Institute of Medical Genetics, University of Oslo, Oslo, Norway

4.1 SUMMARY

Not everybody who is exposed to a harmful environment or life style contracts disease. Evidence is increasing that genetic factors are important in the aetiology of many common disorders, including several diseases that are traditionally thought of as caused by environmental factors alone. Plausibly, environmental or life-style factors preferentially cause disease in those with a genetic predisposition.

Among the common disorders thought to be caused by environmental exposures are several that have a genetic component to their aetiology. Even though ideally any working place should be safe for everybody, it may not always be possible to protect fully those at the highest risk. Accordingly, genetic predisposition to common disorders such as coronary heart disease, several categories of cancers and atopic asthma needs to be considered in efforts to prevent diseases resulting from occupational exposure. Useful markers of coronary heart disease risk already exist, and the event of gene technology holds great promise for more precise genetic prediction of risk for many common disorders.

Voluntary and confidential predictive genetic testing may offer significantly improved possibilities to protect susceptible individuals. The result of predictive genetic tests should be the property of the tested person alone.

4.2 INTRODUCTION

By no means does everybody exposed to an unhealthy environment contract disease. Experimental animal studies have shown that genetic variation in response to several xenobiotics exists in mice and biochemical variants directly or indirectly affect xenobiotics metabolism (such as quantitative differences in the activity of ß-glucuronidase or glutathione reductase). Although the exact reason for differences between individuals in reaction to xenobiotics is in most cases unknown, it is likely that genetic differences between individuals, for example in enzyme functions, are particularly important.

It was the knowledge that genetic differences in reaction to xenobiotics (including drugs) exist that caused the field of ecogenetics to emerge (for review, see Berg, 1979). In our own studies on mutagenicity of body fluids in bacterial test systems, we have observed individual samples that deviate markedly from general trends (Rivrud *et al.*, 1986). Thus, we have seen body fluids from heavy smokers that exhibit a relatively low level of mutagenicity and body fluids from non-smokers that have a quite strong mutagenic effect in such test systems. This illustrates the need to be aware of individual differences in attempting to assess mutagenicity of compounds by comparing exposed and non-exposed humans.

The interaction between chemical exposure and genetic susceptibility to major diseases is still unclear. This paper will focus on genetic susceptibility to common disorders, and three examples will be briefly discussed.

4.3 GENE–ENVIRONMENT INTERACTION AND COMMON DISORDERS

It is a widespread misconception that genetic factors cannot be important in the aetiology of a disorder if it has been proven that environmental, life-style or nutritional factors play a role. It is possible that this misconception is a remnant of an outdated 'nature or nurture' thinking. In reality, there are many examples of disorders where genes as well as environmental factors are aetiologically important.

Another widespread misconception is that there is little or nothing one can do with a disease or with risk factors for the disease if genetic determinants are important. This misconception may result from the situation that prevails for most of the simply inherited disorders of childhood where the therapeutic possibilities are limited. This line of reasoning is not valid for many common disorders where genetic and environmental factors act in concert.

There have been considerable changes in this century in the frequency of atherosclerotic diseases, a group of disorders where genetic factors are known to be important (Berg, 1983). Thus, the frequency of myocardial infarction fell impressively in Norway and other European countries with the greatly changed living conditions of World War II, only to increase again after the war, and the USA and several other countries have experienced a significant decline in myocardial infarction mortality in recent years. Such changes in frequencies cannot be caused by changes in gene frequencies because this would take many generations. The explanation must be that even people with a significant genetic predisposition to coronary heart disease (CHD) may escape disease if life-style, nutritional or other environmental factors are favourably changed.

The unifying concept in cases where there is evidence that genetic as well as environmental factors are of importance in the aetiology of a disorder is that life-style, nutritional or other environmental factors primarily cause disease in people who have some form of genetic predisposition. The realization that genetic factors are of importance in the aetiology of a common disease makes it necessary to try to identify the genes involved. Such attempts would reasonably comprise overt disease as well as risk factors that could perceivably be influenced by genes (such as lipid or apolipoprotein parameters in the case of CHD).

In addition to attempts to uncover associations between genetic markers and absolute risk factor levels, a more dynamic approach has been launched in the area of risk factors for CHD (Berg, 1981, 1984, 1989; Magnus *et al.*, 1981). The rationale for this approach is that genes may not only be important for absolute risk factor levels but also for the frame within which life-style, nutritional or other environmental factors can cause risk factor variation. A method to uncover such genes ('variability genes' as opposed to 'level genes') has been developed (Berg, 1981, 1984, 1989; Magnus *et al.*, 1981). The method is based on the fact that monozygotic (MZ) twin pairs have identical genes. Accordingly, any difference between the two members of an MZ pair in a quantitative parameter must be caused by environmental factors. A permissive variability gene would cause a larger mean within-pair difference in MZ pairs possessing versus those lacking the gene, whereas the opposite effects would be caused by a restrictive variability gene.

4.4 GENE–GENE INTERACTION

In the complex aetiology of common disorders it is not sufficient to consider effects of single genes and single environmental risk factors, or their interaction. Gene–gene interaction could also be operating. With respect to risk factors for CHD, the first example of gene-gene interaction was recently uncovered (Pedersen and Berg, 1989). Inherited isoforms of apolipoprotein E (apoE) are known to influence total and LDL cholesterol levels, people possessing the apoE4 isoform having higher levels than people lacking the isoform. This effect has been shown in several populations. We analysed a population that had been scored with respect to apoE isoforms in a restriction fragment length polymorphism (RFLP) detectable with the restriction enzyme PvuII at the low-density lipoprotein receptor (LDLR) locus and found that the effect of the apoE4 isoform on cholesterol was limited to people who lacked the polymorphic restriction site. Thus, a normal LDLR gene identified by the presence of the PvuII restriction site totally eliminates the effect of the apoE4 allele. A new, independent population sample was recently analysed and the existence of this gene–gene interaction was confirmed (Pedersen and Berg, in preparation). We conclude that predictive testing with respect to apoE isoforms can lead to incorrect risk estimates, unless examination with respect to LDLR variation is also conducted (Pedersen and Berg, 1989).

4.5 THE CANDIDATE GENE APPROACH

Attempts to identify genes of importance in the aetiology of common diseases were previously made with any random genetic marker that might be available. At present, the search for such genes is based on the 'candidate gene approach'. For some common disorders, such as mental illnesses, a very long list of candidate genes may be produced, whereas for other disorders, shorter lists of candidate genes may be visualized. In the case of CHD, a candidate gene could be any gene whose protein product is:

- involved in lipoprotein structure, lipoprotein metabolism or lipid metabolism;
- involved in thrombogenesis, thrombolysis or fibrinolysis;
- involved in regulation of blood flow in coronary arties;
- involved in regulation of blood pressure;
- involved in reverse cholesterol transport;
- present in atherosclerotic lesions;
- involved in the regulation of growth of atheroscerotic lesions;
- involved in the early development of the coronary arteries.

At the level of lipoproteins the candidate gene approach led in the 1970s to the discoveries of close association between genetically determined Lp(a) lipoprotein and CHD, of association between cholesterol level and apoE isoforms and of association between lipid level and the Ag(x) allotypes of low-density lipoprotein (LDL).

The new DNA technology has made it possible to search for genetic predictors of disease risk at many candidate loci. For example, several studies have been conducted on DNA variation at apolipoprotein loci on one hand and CHD risk factors or overt disease on the other (Berg, 1986; Hegele *et al.*, 1986; Law *et al.*, 1986; Ordovas *et al.*, 1986).

4.6 GENETIC PREDISPOSITION TO CORONARY HEART DISEASE

Among the common disorders, the evidence that genetic factors are frequently important is strongest with respect to CHD. Aggregation of CHD cases in families as well as high levels of heritability of several lipid or apolipoprotein parameters had proven that genetic predisposition to CHD exists, before DNA technology became available. Genes are particularly important when CHD occurs at a relatively young age (prior to 55–60 years in males or prior to 60–65 years in females) (for review, see Berg, 1983). Nora and his coworkers (1980) found a heritability of 0.56 for ischaemic heart disease occurring prior to age 55 even after cases of monogenic hyperlipidemia had been excluded, and having a first-degree relative with premature CHD is clearly a risk factor in its own right. With respect to CHD risk factors, level gene as well as variability gene effects have been uncovered (Berg, 1989). Not only DNA variation at apolipoprotein loci seems to affect risk factors but also RFLPs at the locus for cholesteryl ester transfer protein (CETP). This shows the need to study components involved in 'reverse cholesterol transport' in efforts to understand CHD aetiology (Berg *et al.*, 1989; Kondo *et al.*, 1989).

Lp(a) lipoprotein appears to be one of the best predictors of CHD risk (Berg *et al.*, 1974; Rhoads *et al.*, 1986; and Brown and Goldstein, 1987). Rhoads *et al.* (1986) calculated a population attributable risk of 28% for myocardial infarction by age 60, for men with an Lp(a) lipoprotein concentration in the top quartile of the population distribution, and Durrington *et al.* (1988) found that practically all familial clustering of younger CHD cases in the absence of monogenic hyperlipidemia was caused by a high level of Lp(a) lipoprotein.

Although it is generally considered that life-style factors such as smoking, lack of exercise and an unhealthy diet are the most important environmental factors in the aetiology of CHD, there is also evidence that occupational exposure can contribute to CHD risk. Thus, excess mortality from CHD has been found in chimney sweeps (Gustavsson *et al.*, 1987) as well as in workers at a municipal waste incinerator (Gustavsson and Gustavsson, 1988) and in workers in a coke oven plant (Ahlborg *et al.*, 1987). Evidently, more attention should be paid to occupational exposure that may contribute to CHD risk.

Important risk prediction can be derived from family history alone but cholesterol and Lp(a) lipoprotein measurements should be included in predictive CHD risk testing as well as in the diagnostic work-up of CHD cases where the disease has occurred at an early age. Some of the DNA variants at candidate loci (such as the apolipoprotein B (aopB), CETP and LDLR loci) may soon become useful in predictive testing.

Such genetic testing with respect to CHD risk would be most worthwhile since preventive measures are available and should be utilized to the fullest in order to preserve health in people at increased risk because of their genetic constitution.

4.7 GENETIC PREDISPOSITION TO CANCER

Familial aggregation of cases of cancer has been known for a long time. Thus, Waaler (1931), examining the families of 616 males and 786 females with cancer, found increased cancer frequency in the parents, brothers and sisters of cancer patients. By the mid–1970s about 200 conditions known or believed to be monogenic were on record as having neoplasia as a characteristic or as a frequent complication (Mulvihill, 1977). In approximately one-third of the conditions was the mode of inheritance considered definitely proven (Mulvihill, 1977; Berg, 1979). The catalogue of Mendelian traits in man (McKusick, 1988) has continued to grow and now has 4344 items as opposed to 2336 in the mid–1970s. At that time, 8.6% of the entries were conditions that had cancer as a characteristic or frequent complication (Mulvihill 1977; Berg, 1979). Whether or not the percentage of conditions exhibiting propensity to neoplasia is changed, it may safely be assumed that the genetic repertory of human neoplasia is even larger than the impression created by the analyses conducted in the mid–1970s.

The importance of genetic factors in tumour development is, however, not limited to rare, single-gene diseases or heterozygotes for certain rare disorders (such as ataxia telangiectasia). In some instances, frequently occurring tumours (such as breast, lung or colon cancer) exhibit an aggregation of cases in families in a way that is highly suggestive of an effect of genes; shared environment is unlikely to explain fully the aggregation. Empirical knowledge from such aggregations can be used to estimate the risk to relatives for example for breast cancer, given information about bilaterality versus unilaterality and age at onset in affected persons. Bilateral breast cancers occurring before menopause in a mother and a sister constitute a very high risk for breast cancer (Mulvihill, 1977).

Whereas no single genetic marker suitable for general screening for cancer risk is yet available, taking a careful family history and conducting thorough clinical examination can uncover increased cancer risk. The occurrence of cancer in an individual under any of the following circumstances may indicate an increased genetic susceptibility to malignancy (Parry *et al.*, 1987; Berg, 1988) that warrants genetic evaluation:

- cancer in both of a pair of organs that is thought not to be the result of metastases, for example, both breasts, both adrenal glands, both kidneys;
- more than one focus of cancer in a single organ (multicentric tumors), for example, multiple retinoblastomas in one eye;
- two or more distinct cancers (multiple primary malignancies), for example, breast and ovarian cancers, endometrial and colon cancer;
- cancer occurring at an atypical age, at an atypical site or in the sex that is usually less frequently affected, for example, lung cancer in a non-smoking woman;
- cancers associated with other conditions, such as birth defects, one of the 200 single-gene disorders known to be complicated by neoplasia, or with precursor lesions;
- unusual or rare cancers, for example pheochromocytoma.

A family history of cancer may be significant even in evaluating a patient who does not meet any of the above criteria. Two guidelines that incorporate knowledge of the family history have been proposed (Parry *et al.*, 1987; Berg, 1988) for identifying high-risk situations:

- one first-degree relative (brother, sister, parent or offspring) who has cancer and meets any of the criteria above;
- two first-degree relatives with any cancer at a relatively young age or a rare cancer at any age. Further evaluation is not necessary, however, if the sibship is large and the tumour types are common for the sex and age at diagnosis of the patient in question.

Whenever a genetic risk situation is identified, a number of preventive measures including surveillance of high-risk individuals to secure early detection of new or recurrent disease and limiting exposure of high-risk individuals should be considered (Parry *et al.*, 1987; Berg, 1988).

4.8 ASTHMA, ALLERGIES

Cases of atopic asthma and other allergies are known to aggregate in families in a way indicating a strong genetic predisposition. Dominant inheritance of atopy was suggested by family studies of immunoglobulin E (IgE) responses to common, inhaled antigens (Cookson *et al.*, 1989). Close linkage between IgE response and a DNA marker on chromosome 11, confirming a single-gene effect, was recently detected (Cookson and

Hopkin, 1989; Cookson *et al.*, 1989). This discovery makes it likely that more closely linked DNA markers will soon be discovered and that eventually one gene with major effect on the risk of allergy will be cloned. In families, a closely linked gene could be followed to assess the risk of individuals to develop allergy. In the future, gene-specific probes can probably be used for population screening.

Allergies are frequently caused by occupational exposure and in some cases, sufficient reduction of exposure is difficult to achieve. Therefore, it would be in the best interest of those potentially at risk to be protected against such exposure. Thus, it would be very useful if in the future practical possibilities for predictive testing with respect to allergy risk were to become available.

4.9 ETHICAL, LEGAL AND SOCIAL ISSUES IN PREDICTIVE GENETIC TESTING

Predictive genetic testing is to a certain extent already possible with respect to CHD and is likely to become possible with respect to several common disorders in the not too distant future. It is therefore an urgent matter to develop rules that will protect those who want to be tested, against unfavourable use of test results. Tests should be offered on a voluntary basis for the purpose of preventing disease in those who need it the most. It is important that the right to abstain from testing be respected and that no pressure to undergo tests is put on individuals or groups.

For predictive tests to be ethically acceptable, it must be possible to offer some protective advice to people who are at increased risk. It is essential that protective advice be offered at the same time as information about increased risk is given.

If predictive genetic testing is to be widely used to prevent disease, it is of the utmost importance that the public feels that the test results will not be used to its disadvantage. It would be ethically questionable to conduct genetic testing within frameworks of industrial medicine where test results could in principle become available to the employer. Results of predictive genetic tests should be considered as the individual's personal property and it should be explicitly forbidden for insurance companies, employers, pension funds, educational institutions and the military to ask questions about predictive genetic tests or to require that such tests to be performed. The strict rules proposed concerning protection of data should be enforced by explicit laws (Berg and Fletcher, 1986).

ACKNOWLEDGEMENTS

Work in the author's laboratory is supported by the Norwegian Council on Cardiovascular Disease, the Norwegian Research Council for Science and the Humanities and Anders Jahres Foundation for the Promotion of Science.

REFERENCES

Ahlborg, G. Jr, Hillert, L. and Tallgren, U. (1987) Dödsorsaker och cancerincidens bland anställda vid ett koksverk. (Causes of death and cancer incidence among workers at a coke oven plant.) *Hygiea*, **96**, 134.

Berg, K. (1979) Inherited variation in susceptibility and resistance to environmental agents, in *Genetic Damage in Man Caused by Environmental Agents* (ed. K. Berg), Academic Press, New York, pp. 1–25.

Berg, K. (1981) Twin research in coronary heart disease, in *Twin Research 3: Epidemiological and Clinical Studies* (eds L. Gedda, P. Parisi and W. E. Nance), Alan R. Liss, New York, pp. 117–30.

Berg, K. (1983) Genetics of coronary heart disease, in *Progress in Medical Genetics* (eds A. G. Steinberg, A. G. Bearn, A. G. Motulsky and B. Childs), WB Saunders, Philadelphia, pp. 35–90.

Berg, K. (1984) Twin studies of coronary heart disease and its risk factors. *Acta Genet. Med. Gemellol.*, **33**, 349–66.

Berg, K. (1986) DNA polymorphism at the apolipoprotein B locus is associated with lipoprotein level. *Clin. Genet.*, **30**, 515–20.

Berg, K. (1988) Strategies to control cancer through genetics, in *Unusual Occurrences as Clues of Cancer Etiology* (eds R. W. Miller, S. Watanabe, J. F. Fraumeni, *et al.*), Japan Scientific Societies Press, Tokyo, pp. 283–94.

Berg, K. (1990) Level genes and variability genes in the etiology of hyperlipidemia, in *From Phenotype to Gene in Common Disorders* (eds K. Berg, N. Retterstøl, and S. Refsum), Munksgaard, Copenhagen, pp. 77–90.

Berg, K., Dahlén, G. and Frick, M. H. (1974) Lp(a) lipoprotein and pre-β_1-lipoprotein in patients with coronary heart disease. *Clin. Genet.*, **6**, 230–5.

Berg, K. and Fletcher, J. C. (1986) Ethical and legal aspects of predictive testing. *Lancet*, **i**, 1043.

Berg, K., Kondo, I., Drayna, D. and Lawn, R. (1989) 'Variability gene' effect of cholesteryl ester transfer protein (CETP) genes. *Clin. Genet.*, **35**, 437–45.

Brown, M. S. and Goldstein, J. L. (1987) Teaching old dogmas new tricks. *Nature*, **330**, 113–14.

Cookson, W. O. C. M. and Hopkin, J. M. (1990) Towards a gene for atopy and atopic asthma, in *From Phenotype to Gene in Common Disorders* (eds K. Berg, N. Retterstøl and S. Refsum), Munksgaard, Copenhagen, pp. 214–23.

Cookson, W. O. C. M., Sharp, P. A., Faux, J. A. and Hopkin, J. M. (1989) Linkage between immunoglobulin E responses underlying asthma and rhinitis and chromosome 11q. *Lancet*, **i**, 1292–5.

Durrington, P. N., Hunt, L., Ishola, M. *et al.* (1988) Apolipoproteins (a), AI, and B and parental history in men with early onset ischaemic heart disease. *Lancet*, **i**, 1070–3.

Gustavsson, P. and Gustavsson, A. (1988) Dödsorsaker bland arbetare vid en kommunal sopförbränningsanläggning (Causes of death among workers at a municipal waste burning plant). *Arbete och Hälsa*, **18**, 1–21.

Gustavasson, P., Gustavsson, A. and Högstedt, C. (1987) Excess mortality among Swedish chimney sweeps. *Br. J. Ind. Med.*, **44**, 738–43.

Hegele, R. A., Huang, L. S., Herbert, P. N. *et al.* (1986) Apolipoprotein B-gene DNA polymorphisms associated with myocardial infarction. *N. Engl. J. Med.*, **315**, 1509–15.

Kondo, I., Berg, K., Drayna, D. and Lawn, R. (1989) DNA polymorphism at the locus for human cholesteryl ester transfer protein (CETP) is associated with high density lipoprotein cholesterol and apolipoprotein levels. *Clin. Genet.*, **35**, 49–56.

Law, A., Powell, L. M., Brunt, H. *et al.* (1986) Common DNA polymorphism within the coding sequence of the apolipoprotein B gene associated with altered lipid levels. *Lancet*, **i**, 1301–3.

McKusick, V. A. (1988) *Mendelian Inheritance in Man*, 8th edn, Johns Hopkins University Press, Baltimore.

Magnus, P., Berg, K., Børresen, A.-L. and Nance, W. E. (1981) Apparent influence of marker genotypes on variation in serum cholesterol in monozygotic twins. *Clin. Genet.*, **19**, 67–70.

Mulvihill, J. J. (1977) Genetic repertory of human neoplasia, in *Progress in Cancer Research and Therapy* (eds J. J. Mulvihill, R. H. Miller and J. F. Fraumeni Jr), Vol. 3, Raven Press, New York, pp. 137–43.

Nora, J. J., Lortscher, R. H., Spangler, R. D. *et al.* (1980) Genetic-epidemiologic study of early-onset ischemic heart disease. *Circulation*, **61**, 503–8.

Ordovas, J. M., Schaefer, E. J., Salem, D. *et al.* (1986). Apolipoprotein A-I gene polymorphism associated with premature coronary artery disease and familial hypoalphalipoproteinemia. *N. Engl. J. Med.*, **314**, 671–7.

Parry, D. M., Berg, K., Mulvihill, J. J. *et al.* (1987) Strategies for controlling cancer through genetics: report of a workshop. *Am. J. Hum. Genet.*, **41**, 63–9.

Pedersen, J. and Berg, K. (1989) Interaction between low density lipoprotein receptor (LDLR) and apolipoprotein E (apoE) alleles contributes to normal variation in lipid level. *Clin. Genet.*, **35**, 331–7.

Rhoads, G. G., Dahlén, G., Berg, K. *et al.* (1986) Lp(a) lipoprotein as a risk factor for myocardial infarction. *JAMA*, **256**, 2540–4.

Rivrud, G. N., Berg, K., Anderson, D. *et al.* (1986) Study of the amniotic fluids from smokers and nonsmokers in the Ames test. *Mutat. Res.*, **169**, 11–16.

Waaler, G. H. M. (1931) *Über die Erblichkeit des Krebses.* (On the inheritance of cancer.) Bröggers Boktrykkeri AS, Oslo.

PART THREE

Experimental Evidence

5 Evidence in laboratory animal studies of genetic predisposition to effects of toxic chemicals

M. A. Mehlman

Department of Environmental and Community Medicine, Robert Wood Johnson Medical School, Piscataway, NJ, USA

5.1 SUMMARY

In toxicology testing, the choice of animal species and strain is crucial. The significance of genetic differences is illustrated by recent data on the neurotoxicity of MPTP, liver toxicity of acetaminophen and sensitivity to carcinogens. A useful approach to elucidate the variability in genetic predisposition would be to compare several isogenic stocks in chronic toxicity studies. However, the genetic factors are at this time largely unknown.

5.2 INTRODUCTION

Species and strain differences in response to xenobiotics are well known, but the underlying reasons, including genetic factors for the observed differences, are not well understood. A primary reason for the use of animals in toxicological studies is their use as surrogate test models for humans. Humans are genetically diverse, and so are other animals. Attention must therefore be paid to differences between strains and species in selecting the appropriate animal model in order for it to be a useful predictor of human response. Table 5.1 shows interspecific differences in response to some known human carcinogens. Most notable are the tissue specificity and multipotential carcinogenic activity of these chemicals.

Table 5.1 Species differences in response to known human carcinogens[a]

Carcinogen	*Target Organ*	
	Human	Other species
Aflatoxin	Liver	Monkey, Rat, Duck, Fish } Liver
		Rat – Stomach, colon, kidney
		Mouse – Lung
Benzidine	Bladder	Dog – Bladder
		Rat – Liver, colon, zymbal gland
		Mouse – Liver
Vinyl chloride	Liver Lung Brain	Mouse, Rat } Liver, lung, kidney, zymbal gland
		Hamster – Liver
Benzene	Leukaemia	Rat[b] – Mouth, nasal cavity, skin, forestomach, mammary gland, liver, haemolymphoreticular system, zymbal gland

[a]*IARC Monographs*, Suppl. 1, 1979.
[b]Maltoni *et al.*, 1985.

5.3 GENETICS OF LABORATORY ANIMALS

For laboratory studies, three types of animal are used (Festing, 1979):

1. animals from a genetically heterogeneous or nonisogenic group, such as an outbred stock, F_2 hybrid or genetically heterogeneous colony;
2. animals from a genetically uniform or isogenic group, such as inbred strains and F_1 hybrids;
3. animals expressing a major gene mutation.

Isogenic animals are genetically identical. An inbred strain is one derived from a single breeding pair in the twentieth or a subsequent generation of brother–sister mating, or its equivalent, and its main genetic properties are that all members are genetically identical and homozygous at all loci. An F_1 hybrid is the first-generation cross of two inbred strains. Members of an F_1 hybrid are isogenic but heterozygous at all loci at which the parental strains differ and therefore will not breed true. A non-isogenic stock is one which does not fit the criteria for an isogenic stock. An outbred stock is a stock of animals usually maintained as a closed colony, with either random mating or a mating system designed to minimize inbreeding (Festing, 1979). Common outbred stocks are Wistar, Sprague-Dawley and Long Evans rats; Swiss, Swiss-Webster,

CD–1, CFW and CF–1 mice; and Duncan-Hartley guinea pigs. Inbred stocks include F344 and OM rats and C57BL/6J, DBA/2J and BALB/C mice (Taylor, 1972).

Table 5.2 Properties of isogenic strains and non-isogenic stocks for consideration in the choice of animal for toxicological studies (from Festing, 1979)

Property	*Isogenic*		*Non-isogenic*	
	Inbred strain	*F_1 hybrid*	*Outbred*	*F_2 or heterogeneous*
Isogenicity	Very high	Very high	Intermediate to low	Very low
Long-term genetic stability	High	Very high	Low	Low
Phenotypic	Intermediate to high	Very high	Intermediate	Low
International distribution	Widespread	Widespread	Restricted	Very restricted
Background data	Much	Intermediate	Intermediate	Little

5.4 CHOICE OF ANIMAL STRAIN FOR TOXICOLOGICAL STUDIES

The choice of animal species and strain is extremely critical in chronic bioassay studies for detecting carcinogenic activity. Among general considerations cited in selecting the proper species and strain are spontaneous tumour incidence, sensitivity, hardiness and longevity (Sontag *et al.*, 1976). A summary of properties of isogenic and non-isogenic stocks that need to be considered for long-term studies is shown in Table 5.2. The Cancer Bioassay Programme of the NCI uses primarily B6C3F_1 hybrid mice and either the F344 or OM rat. These are isogenic strains with high long-term genetic stability, high phenotypic uniformity, widespread international distribution and a fair amount of historical data. The pros for using isogenic stocks include the following:

1. minimal variability, which increases the precision of statistical analyses, thus allowing better detection of differences between treated and control groups;
2. high response since the strain may be sensitive towards the chemical;
3. good repeatability of studies.

On the other hand, a non-isogenic stock might be a better model since it exhibits a wider range of susceptibilities, more in keeping with the heterogeneous human population, and it may be capable of a toxic response not possible in an isogenic strain. An isogenic strain, however, may require large doses for effects to be seen and problems may be encountered with repeatability of the results (Littlefield and Kodell, 1979).

Another approach for chronic toxicity testing would be the use of several isogenic stocks in a factorial experimental design (Heston *et al.*, 1973; Festing, 1975). Such an experimental design would provide a wider basis for extrapolation to humans, has a higher repeatability, and might be used to show that a response is genetically determined (Festing, 1979). Some examples will be discussed below to illustrate the considerations involved.

5.4.1 Sensitivity to a neurotoxin in different strains of mice

The compound MPTP (1-methyl–4-phenyl–1,2,3,6-tetrahydropyridine) has been linked with Parkinsonian-like symptoms in several young drug abusers (Davis *et al.*, 1979; Langston *et al.*, 1983). This compound is metabolized by monoamine oxidase B (MAO-B) to 1-methyl–4-phenyl-dihydropyridinium ($MPDP^+$), which is rapidly converted to the neurotoxic species 1-methyl–4-phenyl-pyridinium ion (MPP^+). It has been suggested that MPTP-induced neurotoxicity is linked to the formation of the toxic species MPP^+ via the monoamine oxidase pathway and the accumulation of this toxic species in dopaminergic neurons by the dopamine transport system.

In studies on the influence of dose and dosing interval on MPTP-induced dopaminergic neurotoxicity in mice, Sonsalla and Heikkila (1986) found that the toxic effects of MPTP (4×20 mg/kg injection, 2-hour intervals), using as endpoints mortality and depletion of dopamine and its metabolites in the neostriatum, were greatest in C57BL mice and diminished in CF-W (Swiss-Webster) and BALB/C mice. No significant effects were observed in CF–1 or CD–1 mice.

These differences could not be accounted for by differences in MAO-B activity or differences in the dopamine transport system between the various strains of mice. However, substantial differences in brain levels of MAO-B-catalyzed MPTP oxidation products were found following MPTP administration. The highest metabolite levels were found in the most MPTP-sensitive BC57BL mouse strain, and lowest metabolite levels in CD–1 mice, the least MPTP-sensitive strain examined (Table 5.3).

Table 5.3 Sensitivity towards MPTP and brain levels[a] of MPTP and its metabolites in different strains of mice[b,c]

Strain	*MPTP*	*MPDP⁺ MPP⁺*	*% Mortality*[d,e]
C57BL	7.3±5.9	14.8±1.0	60
CF-W	5.6±2.9	10.6±2.0	10
CD–1	3.8±0.8	8.8±2.6	0

[a]Expressed in pmole/mg tissue.
[b]From Sonsalla and Heikkila, 1988.
[c]One subcutaneous injection of MPTP (20 mg/kg; 10μCi ^{3}H-MPTP).
[d]From Sonsalla and Heikkila, 1986.
[e]After MPTP administration, 4 × 20 mg/kg at 2-hour intervals.

These results support the hypothesis that the formation and/or concentration of the MPP^+ species in the brain are important determinants of neurotoxicity. These studies point to differences in metabolism to a toxic species as a determining factor of strain sensitivity.

5.4.2 Strain differences in acetaminophen toxicity in normal and diabetic rats

Studies by Price and Jollow (1986) have shown that normal male Sprague-Dawley (SD) rats are more resistant to acetaminophen-induced liver necrosis compared with normal male Long-Evans Hooded (LEH) rats. The greater resistance towards acetaminophen (AA) toxicity in SD rats was associated with a faster elimination of AA via metabolism to non-toxic glucuronide and sulfate conjugates. The difference in glucuronidation capacities between normal SD and LEH rats was not due to differences in either the amount of glucuronyl transferase or basal hepatic levels of the cofactor UDPGA.

When SD and LEH rats were treated with streptozotozin, the diabetic state protected LEH but not SD rats from AA-induced liver necrosis. Although enhanced elimination of conjugates was found in diabetic animals of both strains, LEH diabetic rats had both enhanced sulfation and glucuronidation of AA metabolites compared with only enhanced sulfation in SD diabetic rats.

The authors concluded that a major difference in susceptibility between SD and LEH rats towards AA toxicity appears to be due to the capacity to clear the drug through non-toxic pathways. In comparison with the strain differences to MPTP toxicity in mice that appear to be related to greater metabolism to the neurotoxic intermediate in the more MPTP-susceptibile strain, the results of the AA studies point to strain differences in detoxification pathways.

5.4.3 Sensitivity to complete and two-stage carcinogenesis in different strains and stocks of mice

Complete carcinogens are agents that can induce tumours without exposure to a second agent. Tumour promoters, on the other hand, are agents that promote the expression of the tumorigenic potential of a subcarcinogenic exposure to a complete carcinogen. The tumorigenic response of mice to the complete carcinogen dimethylbenz(a)anthracene (DMBA) varies considerably with the strain of mouse used in the bioassay (Table 5.4). Using susceptibility to tumour induction by the complete carcinogens benzo(a)pyrene (B(a)P) and DMBA, mouse stocks and strains have been ranked in decreasing order of susceptibility as follows:

SENCAR>CD–1>C57BL/6>ICR/Ha>Swiss>C3H.

Sensitivity with respect to tumour induction using the initiation-promotion protocol decreases in the order:

SENCAR > CD–1 > ICR/Ha > Swiss > BALB/c > C57BL/6 > CeH > DBA/2 (Reiners *et al.*, 1983).

Current theory suggests that binding of metabolites of carcinogens to DNA plays a critical role in cancer induction by complete carcinogens. Although the amount of polycyclic aromatic hydrocarbon bound to DNA correlates well with the ability of polycyclics to induce skin tumours in any individual mouse stock or strain (Phillips *et al.*, 1979; Slaga *et al.*, 1980), the correlation does not appear to hold in comparisons between mouse strains and tumour response to complete carcinogens (Phillips *et al.*, 1978).

The process of tumour induction by complete carcinogens is believed to consist of both an initiation and a promotion phase. To account for differences in strain sensitivity towards complete carcinogens, differences in the promotional phase of complete carcinogenesis have been suggested. With respect to differences towards two-stage carcinogenesis and complete carcinogenesis in the same mouse strain, differences in the process of promotion between the complete and two-stage protocol have been suggested as a possible explanation (Reiners *et al.*, 1983).

5.5 CONCLUSIONS

As indicated by the above examples, genetic differences exist between different species and different strains of animals. However, the underlying genetic factors responsible for differences in susceptibility to toxic

Table 5.4 Differences in DMBA-induced tumours in six strains of mice[a,b]

Strain	*No. of tumour-bearing animals/total animals*	*Total No. of tumours*	*Papillomas*	*No. of histologically verified tumours* Kerato acanthomas	*Squamous cell carcinomas*	*Other*
Swiss	29/40	47	23	11	11	2
BALB/C	40/40	96	47	21	25	3
A	30/40	55	27	11	16	1
C57BL	23/40	31	14	7	10	–

[a]From Stenback, 1980.
[b]10 μg DMBA 2 × week.

chemicals are largely unknown. In the words of Festing and Eydman (1987) 'most strain differences are under polygenic control, and so far there have been no successful attempts to isolate individual loci'.

REFERENCES

Davis, G. C., Williams, A. C., Markey, S. P., *et al* (1979) Chronic parkinsonism secondary to intravenous injection of meperidine analogues. *Psychiatry Research*, **1**, 249–54.

Festing, M. F. W. (1975) A case for using inbred strains of laboratory animals in evaluating the safety of drugs. *Food Cosmet. Toxicol.*, **13**, 369–75.

Festing, M. F. W. (1979) Properties of inbred strains and outbred stocks with special reference to toxicity testing. *J. Toxicol. Environ. Health*, **5**, 53–68.

Festing, M. F. W. and Eydman, T. (1987) Genetic variation in the response of mice to xenobiotics *in vitro*. General methodology and response to some model compounds. *Toxic. in Vitro*, **1**, 115–25.

Heston, W. E., Desmukes, B. and Vlahakis, G. (1973) Effects of the antifertility drug Enovid in five strains of mice with particular regard to carcinogenesis. *J. Natl. Cancer Inst.*, **51**, 209–24.

International Agency for Research on Cancer (1979) *Chemicals and Industrial Processes Associated with Cancer in Humans*, IARC Monographs, Vol. 1–20, IARC Monographs Suppl. 1, IARC, Lyon.

Langston, J. W., Ballard, P., Tertrud, J. W. and Irwin, I. (1983) Chronic Parkinsonism in humans due to a product of meperidine-analog synthesis. *Science*, **219**, 979–80.

Littlefield, N. A. and Kodell, R. A. (1979) Influence of genetic population structure on the results of chronic toxicity studies. *J. Toxicol. Environ. Health*, **5**, 121–9.

Maltoni, C., Conti, B., Cotti, G. and Belpoggi, F. (1985) Experimental studies on benzene carcinogenicity at the Bologna Institute of Oncology: current results and ongoing research, in *Benzene, Scientific Update* (ed. M. A. Mehlman), Alan R. Liss, New York, pp. 415–46.

Phillips, D. H., Grover, P. L. and Sims, P. (1978) The covalent binding of polycyclic hydrocarbons to DNA in the skin of mice of different strains. *Int. J. Cancer* **22**,487–494.

Phillips, D. H., Grover, P. L. and Sims, P. (1979) A quantitative determination of the covalent binding of a series of polycyclic hydrocarbons to DNA in mouse skin. *Int. J. Cancer*, **23**, 201–8.

Price, V. F. and Jollow, D. J. (1986) Strain differences in susceptibility of normal and diabetic rats to acetaminophen heptotoxicity. *Biochem. Pharmacol.*, **35**, 687–95.

Reiners, J. Jr, Davidson, K., Nelson, K. *et al.* (1983) Skin tumour promotion: a comparative study of several stocks and strains of mice, in: *Organ and Species Specificity in Chemical Carcinogenesis* (eds R. Langenbach, S. Nesnow and J. M. Rice), Plenum Press, New York and London, pp. 173–86.

Slaga, T. J., Fischer, S. M., Weeks, C. E. and Klein-Szanto, A. J. P. (1980) Multistage chemical carcinogenesis in mouse skin, in *Biochemistry of Normal and Abnormal Epidermal Differentiation* (eds I. A. Bernstein and M. Seiji), University of Tokyo Press, Tokyo.

Sonsalla, P. K. and Heikkila, R. E. (1986) The influence of dose and dosing on MPTP-induced dopaminergic neurotoxicity in mice. *Eur. J. Pharmacol.*, **129**, 339–45.

Sonsalla, P. K. and Heikkila, R. E. (1988) Neurotropic effects of 1-methyl-4-phenyl-1,2,3,6-tetrahydroxypyridine (MPTP) and methamphetamine in several strains of mice. *Prog. Neuro. Psychopharmacol. Biol. Psychiat.*, **12**, 345–540.

Sontag, J. M., Page, N. P. and Saffiotti, M. D. (1976) *Guidelines for Carcinogen Bioassay in Small Rodents*, DHEW Publ. NIH, 76–801.

Stenback. F. (1980) Skin carcinogenesis as a model system: observations on species, strain and tissue sensitivity to 7,12-dimethylbenz(a)anthracene with or without promotion from croton oil. *Acta Pharmacol. Toxicol.*, **46**, 89–97.

Taylor, B. A. (1972) Genetic relationships between inbred strains of mice. *J. Hered.*, **63**, 83–6.

6 Animal models for studies on genetic predisposition to adverse effects of chemical exposure

S. R. Kleeberger and R. C. Levitt

Department of Environmental Health Sciences, Department of Anesthesiology and Critical Care Medicine, The Johns Hopkins Medical Institutions, Baltimore, MD, USA

6.1 SUMMARY

Susceptibility to the toxic and carcinogenic effects of environmental pollutants and drugs varies widely between individuals. Genetic background plays a role in this differential susceptibility, and the need for predictive animal models of genetic susceptibility is becoming increasingly clear. However, the development of well-characterized genetic models has not progressed very rapidly. We have examined some of the existing genetic models, including Mendelian and non-Mendelian loci, for biological responses to a number of pharmacological and environmental agents. A murine model of genetic susceptibility to ozone-induced inflammation is also described, and it serves to illustrate the application of simple genetic techniques for investigating the influence genetic background may have on lung responsiveness to environmental pollutants. Single-gene or Mendelian models have certain advantages with respect to testing hypotheses about the biochemical or cellular mechanisms of gene action, and thus provide a potential understanding of interindividual biological variability in responsiveness to environmental chemicals. Thus, animal models of genetic susceptibility should be further developed and close interaction between the predictive capabilities of the models and human studies should be encouraged.

6.2 INTRODUCTION

Genetic susceptibility to environmental agents in humans has been well documented through population, family and twin studies (Motulsky, 1977; Idle and Smith, 1980; McKusick, 1983; Calabrese, 1984; Festing, 1987; Koenig and Omenn, 1988). However, because of obvious ethical considerations, dose-response relationships and the biochemical and histological mechanisms of the responses to environmental chemicals have been difficult to determine. For the purpose of identification of genetically hypersusceptible individuals who may be at risk by exposure to environmental chemicals, a clear need for predictive animal models of genetic predisposition to adverse effects of these substances for qualitative risk assessment (QRA) in humans has long been recognized (Brieger, 1963; Calabrese, 1978; Mohn, 1978; Gardner, 1979; Festing, 1987). However, surprisingly few well-defined animal models of genetic predisposition currently exist.

Toxicological testing of new drugs and chemicals has led to the development of numerous animal models of QRA for humans, but concern has been expressed about extrapolation from animal to man (Menzel, 1979; Crouch, 1983a,b; Hogan, 1983; Wilson and Crouch, 1984; Clayson, 1985; Hart and Fishbein, 1985; Festing, 1987; Brown *et al.*, 1988; Gregory, 1988). Additionally, because most experimental designs in toxicological testing do not account or control for genetic background, knowledge about the potential genetic contribution to susceptibility or predisposition to responses is generally not obtainable (Festing, 1987). We will briefly describe some of the existing models of genetic predisposition to toxic chemicals, and evaluate the predictive capabilities of the models for the purpose of QRA in humans.

6.3 REVIEW OF GENETIC MODELS

Susceptibility to the toxic and carcinogenic effects of environmental pollutants varies widely between humans (Motulsky, 1978; Omenn, 1978; Price-Evans, 1980). Numerous factors have been suggested to contribute to differential susceptibility, including diet, age, gender, socioeconomic status and hereditary background. This section will be restricted to examination of models of genetic predisposition to toxic effects of environmental chemicals. A number of models of susceptibility to the carcinogenic effects of chemicals also exist (e.g. Gould, 1986), and have been reviewed (Hansen and Cavenee, 1987; Shaikh *et al.*, 1988). The numerous advantages of highly inbred, genetically standardized animals for genetic analyses of biological processes are well known (Festing, 1979, 1987; Taylor, 1980; Green, 1981; Briles *et al.*, 1986) and, consequently, a majority of the work on genetic predisposition to chemicals has been done on inbred strains, largely mice and rats (Festing, 1987). For example, inbred mice have been used extensively in the genetic analyses of immune responses and disease resistance, including lipopolysaccharide-induced interleukin–1 production (Brandwein *et al.*, 1987), and genetic control of resistance to *Listeria* (Cheers *et al.*, 1980; Gervais *et al.*, 1984), *Salmonella* (O'Brien *et al.*, 1980; Plant *et al.*, 1982), *Leishmania* (O'Brien *et al.*, 1980; Plant *et al.*, 1982), amyloidosis (Mortensen *et al.*, 1985) and *Mycobacterium* (Skamene *et al.*, 1982; Skamene, 1986).

The first step in elucidation of the genetic control of a biological process is to screen a series of inbred strains for their specific responses to the agent under study (Festing, 1979). Interstrain variation in responses to xenobiotics has been demonstrated in nearly every study which has made the comparisons (Festing, 1987). These comparisons indicate that the response(s) of these animals may be under strong genetic control, although the mode of inheritance cannot be certain until further genetic analyses are completed. These analyses include breeding studies in which F_1, F_2 and back-cross progeny from phenotypically distinct inbred parental strains are phenotyped for the trait of interest (Festing, 1979). Most studies, however, have suggested that susceptibility to xenobiotics (as well as many common diseases) results from polygenic mechanisms (Festing, 1987; Koenig and Omenn, 1988), and only in a few instances has Mendelian inheritance been demonstrated.

6.3.1 Mendelian and non-Mendelian responses to environmental chemicals

(a) Non-Mendelian (polygenic) responses

As indicated above, the observed genetic interindividual variation in response to many drugs and environmental agents appears to have a polygenic mode of inheritance. In human disorders, polygenic control may be inferred if the trait in question does not appear to conform to predictions of Mendelian inheritance in familial analyses. However, the analysis of a disorder is often complicated by interaction between environmental and genetic factors, and a trait may be classified as polygenic only because Mendelian inheritance is masked by this interaction (Festing, 1987; Koenig and Omenn, 1988). A discussion of this problem in the analysis of genetic disorders and examples of responses which are presumably under polygenic control may be found in Festing (1987).

(b) Mendelian responses

Reviews of animal studies which have examined the influence of genetic factors in toxicological testing reveal few studies of Mendelian loci which have been identified and characterized for the control of responses to environmental chemicals (Kalow, 1962; Meier, 1963; Lush, 1981; Festing, 1987). That is, although strain screen studies have demonstrated that physiological responses to many xenobiotics are under strong genetic control, it appears that most of the responses are influenced by polygenic rather than single-gene inheritance. Four examples of well-defined Mendelian loci for control of responses to xenobiotics are listed below.

The murine Ah *locus*

The *Ah* locus is perhaps one of the most widely recognized and studied loci in pharmaco- and ecogenetics. The inducibility of hepatic cytochrome P_1–450 and associated monooxygenase activity (including aryl hydrocarbon hydroxylase (AHH)) by 3-methylcholanthrene (3-MC) and other polycyclic aromatic hydrocarbons is presumably under the control of this locus. In approximately half of the inbred strains examined (such as C57BL/6), AHH is inducible by 3-MC, while the remainder (for example DBA/2J) are non-inducible (Omenn and Motulsky, 1978). As a result of breeding studies between B6 and D2 strains of mice, a single gene was proposed (*Ah*) to control inducibility of this and associated monooxygenase activities.

Since these studies were first presented (Nebert and Gielen, 1972; Thomas *et al.*, 1972), the genetics of the *Ah* locus have proved to be much more complex than earlier believed. Thus although the differences in regulation of responsiveness between C57BL/6 and DBA/2 can be almost completely explained by the *Ah* locus, several alleles at more than one locus are probably necessary to explain all existing data in the mouse (Nebert and Atlas, 1978). Because of the complexity of this locus (or loci) mapping of the gene has been difficult. However, a report by Poland *et al.* (1987) indicates a map assignment of an *Ah* allele to chromosome 12. The *Ah* locus appears to be the structural gene for a cytosolic protein receptor which combines with polycyclic aromatic inducers to regulate the formation of membrane-bound cytochromes, and it is hypothesized that an altered receptor with diminished affinity for inducers may be the primary defect in non-responsive or non-inducible mice (Poland *et al.*, 1974, 1987; Nebert, 1980; Green, 1981).

The evidence for similar genetic control of AHH inducibility in humans, and its importance in carcinoma prediction, is controversial. Early reports (for example, Kellerman *et al.*, 1973a,b) examined AHH inducibility in cultured mitogen-activated lymphocytes by 3-MC in healthy human subjects and found the distribution of inducibilities to be trimodal, such that low, intermediate and high phenotypes were identified. The data from these studies conformed to the hypothesis that inducibility of AHH resulted from two alleles at a single locus with the three groups representing homozygous low and high alleles and the intermediate heterozygote. A comparison was then made between AHH inducibility and the incidence or risk of bronchogenic carcinoma. The authors of the study concluded that with respect to developing carcinoma, individuals with the high inducibility phenotype were 36 times more likely and individuals with the intermediate phenotype were 16 times more likely than low inducers. However, subsequent studies which have sought to reproduce these findings have been equivocal, and explanations for the inconsistency of the findings have been detailed elsewhere (Nebert *et al.*, 1978; Calabrese, 1984). While the genetic basis for susceptibility to environmentally induced lung cancer and toxicity is well characterized in animal models, a precise and reliable marker or predictor of risk to lung cancer, such as AHH inducibility, has yet to be identified (Calabrese, 1984).

The murine cadmium resistance locus, cdm

Awareness and concern have been growing about the presence of heavy metals such as cadmium in the environment. These metals accumulate in various tissue types and are eliminated slowly, if at all (Nordberg and Kjellstrom, 1979). The effects of heavy metals on man and other animals are not well understood, but they may be involved in kidney damage and cardiovascular disease as well as other disease processes (Fox, 1979). Sensitivity to cadmium-induced necrosis of the testis has been demonstrated to vary greatly between inbred strains of mice such that AKR/J, CBA/J, C57L/J, DBA/2J and SWR/J strains, for example, are susceptible and most others (e.g. A/J, BALB/cJ, C3H/HeJ, C57BL/6J) are resistant (Taylor *et al.*, 1973). Genetic analyses with DBA/2J and C57BL/6J strains demonstrated conclusively that resistance to cadmium-induced testicular necrosis in inbred mice is controlled by a single autosomal recessive gene (*cdm*). Further genetic analyses with these strains established linkage of the *cdm* locus on mouse chromosome 12 (Taylor, 1976).

The murine warfarin resistance locus, War

The blood anticoagulant, warfarin, has been used extensively as a therapeutic agent, and has also been used as a rodenticide. Recently, reports of adverse effects of warfarin treatments have been published and it has been suggested that these effects may be in part related to differential genetic predisposition (Pauli *et al.*, 1987). Differential susceptibility to the toxic effects of warfarin in wild populations of mice and rats has also been reported (Wallace and MacSwiney, 1976; Greaves and Ayres,

1977). A series of breeding experiments was performed by Wallace and MacSwiney (1976) on wild-trapped mice which demonstrated the existence of a major gene for warfarin resistance in the mouse which is dominant in females and semidominant in males, i.e. modifying genetic factors affect the degree of resistance in males but not in females (Green, 1981). These studies also established linkage to a colour mutation, extreme dilution (c^e), on mouse chromosome number 7. A similar mutation for differential resistance to warfarin has been described in the rat, and the mutation occupies an analogous position in that species (Green, 1981).

The acetylator phenotype

The hereditary polymorphic *N*-acetylation of arylamine-containing drugs has been recognized since administration of isoniazid (INH) for tuberculosis began approximately 30 years ago (Weber, 1984; Clark, 1985). The trait has significant effects on the ability to metabolize many drugs such as INH, hydralazine, procainamide, dapsone, aminoglutethimide and various sulfonamides, as well as some environmental chemicals that contain an amine or hydrazine group (Weber, 1984). Individuals exhibit differential drug toxicity as a function of acetylator phenotype, such that slow acetylators may have higher serum concentrations or longer half-life of INH or other drugs. Compared to fast acetylators, slow acetylators have increased risk of peripheral neuropathy from INH treatment (Clark, 1985), systemic lupus erythematosus from hydrazaline (Perry *et al.*, 1970) and procainamide (Henningsen *et al.*, 1975; Woosley *et al.*, 1978), and they may have increased susceptibility to phenytoin and carbamazepine toxicity if receiving INH (Clark, 1985). Hereditary acetylation polymorphisms are known in inbred rabbits (Weber, 1984) and hamsters (Hein *et al.*, 1982), and the acetylator phenotype has been demonstrated to have Mendelian inheritance in inbred strains of mice (Tannen and Weber, 1980). Because of the close resemblance of these genetic models to the human acetylation phenotype, these animal models have proven to be extremely useful for investigating the mechanisms and biological consequences of these polymorphisms (Weber, 1984).

6.4 A MODEL OF GENETIC SUSCEPTIBILITY TO OZONE-INDUCED INFLAMMATION

In recent years, ozone has become a common air pollutant in urban and suburban areas, as well as in the work environment. It is a highly reactive oxidant, and because of its potentially toxic effects, ozone has received increasing attention regarding effects on human health, especially pulmonary function (Menzel, 1984; Mustafa and Tierney, 1978).

Acute exposures to ozone in resting and exercising humans cause increased airway resistance and respiratory rate, as well as decreased forced vital capacity, total lung capacity and tidal volume (Bates, 1972; Folinsbee *et al.*, 1984). Exposure to ozone in humans (Seltzer *et al.*, 1986) and experimental animal models (Kleeberger *et al.*, 1988) also elicits the release of bronchoactive chemical mediators and induces inflammation, a process which has been intimately linked with the pathogenesis of a host of airway diseases, including asthma and chronic obstructive pulmonary disease (Hargreave *et al.*, 1986; Vane and Botting, 1987). Although inflammation is implicated in many disease states, the molecular mechanisms involved in the development of inflammation are not clear (Melmon and Bourne, 1974), and studies designed to investigate these mechanisms are needed (NIH, 1985).

Evidence for the explicit role of genetic susceptability to environmental oxidants has accumulated (Omenn and Motulsky, 1978; Koenig and Omenn, 1988). McDonnell *et al.* (1985) demonstrated consistent inter-subject differences in the magnitude of change in forced vital capacity (FVC) and forced expiratory volume in one second (FEV_1) for ozone exposures greater than or equal to 0.18 ppm, and concluded that the acute responsiveness to ozone has a heritable component. Early work by Goldstein *et al.* (1973) demonstrated great interstrain differences in the lethal effects of ozone on mice, and Costa *et al.* (1985) found differential susceptibility to the effects of ozone in rats. These authors suggested that susceptibility to the toxic effects of ozone may have a genetic basis. Although it is not clear whether differences in intrinsic responsiveness to acute ozone exposures have a bearing on possible subacute or chronic effects in humans (McDonnell *et al.*, 1985), animal models for investigating potential manifestations, including acute and chronic airway inflammation, would be useful.

In our laboratories we have begun to define a genetic model in order to investigate mechanisms of oxidant-induced inflammation and susceptibility which may contribute to the inflammation. In the development of this model we have used bronchoalveolar lavage (BAL) to monitor the infiltration of polymorphonuclear leukocytes (PMNs) into the airways, a classical indicator of inflammation. The experimental protocol entailed exposure of highly inbred strains of mice (male, 5–7 weeks) to 2 ppm ozone or air for 3 hours; pulmonary inflammation was assessed 2, 6 and 24 hours after exposure.

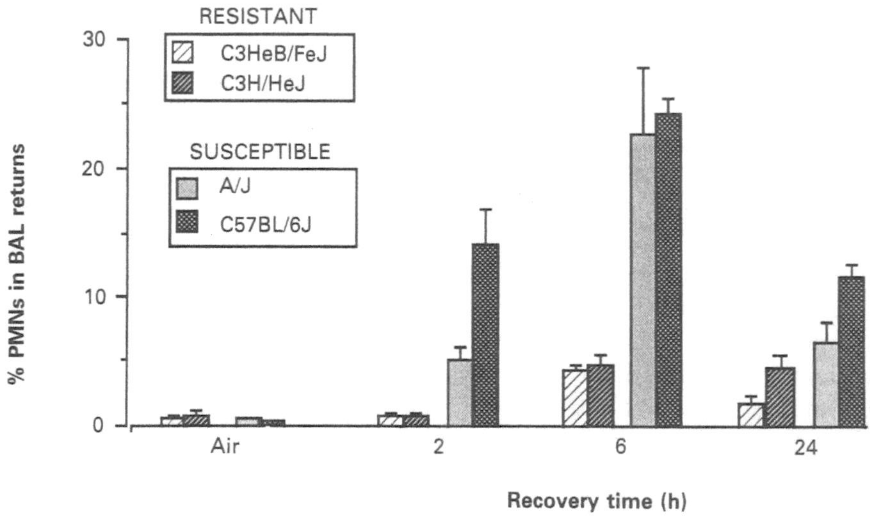

Figure 6.1 The percentage of polymorphonuclear leukocytes (PMNs) in total cells recovered by bronchoalveolar lavage 2, 6 and 24 hours after 3-hour exposure to 2 ppm ozone in C3HeB/FeJ, C3H/HeJ, A/J and C57BL/6J strains of mice. Means ±1 SEM are presented.

6.4.1 Interstrain variation in the inflammatory response to ozone

We have characterized the time course and magnitude of the ozone-induced inflammatory response in 12 highly inbred, genetically standardized strains of mice (Kleeberger *et al.*, 1990). Although the time course (a qualitative parameter) of the response was generally similar between the strains, the magnitude of the response (a quantitative parameter) was highly divergent. The inflammatory response for 4 of the 12 strains is illustrated in Fig. 6.1. The maximum change in PMNs was detected 6 hours following ozone exposure, and recovery towards baseline was evident after 24 hours. The air-exposed (control) animals exhibited no detectable changes in PMNs at any time point when compared to unexposed animals and the data were thus not pooled (Fig. 6.1). The phenotypes of the C57BL/6J (B6, termed SUSCEPTIBLE) and C3H/HeJ (C3, RESISTANT) strains were easily distinguished on the basis of their inflammatory response to O_3. A 20-fold difference in %PMNs was detected between the two strains 2 hours after ozone ($P < 0.001$), and a 5-fold difference was found 6 hours post-exposure ($P < 0.001$; Fig. 6.1). A 2.5-fold difference was still detectable 24 hours after ozone.

The time course and magnitude of the total number of PMNs recovered by BAL were similar to the changes in %PMNs: a 22-fold difference in PMNs was detected between the SUSCEPTIBLE B6 and RESISTANT C3 strains 2 hours after ozone (4.23 ± 1.01 vs $0.19 \pm 0.11 \times 10^4$ PMNs/ml; $P < 0.001$); a 6-fold difference was found 6 hours post-exposure ($P < 0.001$; Table 6.1); and a 5-fold difference was detectable 24 hours after ozone exposure (3.15 ± 0.04 vs $0.63 \pm 0.01 \times 10^4$ PMNs/ml; $P < 0.001$). No detectable changes in the numbers of recovered alveolar macrophages, lymphocytes or epithelial cells were detected at these time points (data not shown).

These data indicated that the magnitude of the acute ozone-induced influx of PMNs exhibited large variation between highly inbred strains of mice, and suggested strong genetic regulation of the response. We have also begun to examine whether interstrain differences in inflammation would be observed in response to non-specific stimuli. In C3 and B6 mice, peritoneal inflammation was induced by IP injection of 10% glycogen, and after 5 hours cells were harvested by IP lavage. In seven experiments ($n = 5$/strain per experiment) the %PMNs recruited was not significantly different between strains (C3: $89.3 \pm 1.8\%$ vs B6: $89.4 \pm 1.3\%$), and total numbers of PMNs were also not significantly different (10.9 ± 2.2 vs $7.2 \pm 1.8 \times 10^6$). Results thus indicate that the IP inflammatory response to a non-specific stimulus did not differ between the C3 and B6 strains, and may suggest that the observed difference in ozone-induced inflammation is due to a specific defect related to oxidant injury in one of the strains.

6.4.2 Genetic analysis of the inflammatory response to ozone

To obtain further insight into the apparent genetic regulation of susceptibility to acute ozone-induced pulmonary inflammation, the inflammatory response of the first filial generation (F_1: B6C3F1/J, C3B6F1), back-cross progeny from the RESISTANT C3 and SUSCEPTIBLE B6 strains and the F_2 generation were examined (Kleeberger *et al.*, 1989, 1990). The 6-hour time point was chosen for comparison of changes in PMNs, as earlier studies demonstrated maximum PMN influx at this time.

The numbers and percentages of PMNs in BAL fluid of both F_1 hybrids were not significantly different from those of the RESISTANT C3 strain following exposure to ozone (Table 6.1). The data are consistent with the hypothesis that ozone-induced inflammation in B6 and C3 strains of mice and their progeny is influenced by an autosomal recessive gene. If autosomal recessive inheritance is assumed for the SUSCEPTIBLE phenotype exhibited by the B6 strain, then Mendelian genetics predict that 100% of the back-cross progeny of the F_1 to the C3 strain should exhibit the RESISTANT phenotype. Progeny of the back-cross to the B6 strain should exhibit a 1:1 ratio of RESISTANT to SUSCEP-

Table 6.1 Changes in lavageable polymorphonuclear leukocytes (PMNs) expressed as number and percentage of total recovered cells in C57BL/6J (B6) C3H/HeJ (C3) mice, and their progeny, 6 hours after 3-hour exposure to 2 ppm ozone or filtered air

		Number[b] of PMNs ($\times 10^4$)			*Percentage of PMNs*		
		Air	*Ozone*		*Air*	*Ozone*	
Strain or offspring	*Inf* phenotype[a]	*Mean± 1 SE*	*Mean ± 1 SE*	*Range*	*Mean ± 1 SE*	*Mean ± 1 SE*	*Range*
C57BL/6J	SUSCEPTIBLE	0.09 ± 0.02 (15)	*6.68 ± 0.66 (47)	2.25 - 19.52	0.3 ± 0.1	*24.3 ± 1.2	11.3 - 52.3
C3H/HeJ	RESISTANT	0.03 ± 0.01 (17)	1.05 ± 0.25 (14)	0.23 - 2.18	0.8 ± 0.3	4.7 ± 0.7	1.0 - 9.7
B6C3Fl/J	RESISTANT	0.03 ± 0.02 (5)	1.06 ± 0.32 (10)	0.26 - 2.09	0.2 ± 0.1	3.1 ± 0.5	0.7 - 4.7
C3B6Fl	RESISTANT	0.05 ± 0.02 (5)	1.00 ± 0.27 (15)	0.19 - 2.00	0.4 ± 0.1	2.5 ± 0.4	0.0 - 7.0
C3H/HeJ:BX	RESISTANT	0.03 ± 0.03 (4)	1.55 ± 0.30 (8)	0.41 - 2.92	0.2 ± 0.2	5.3 ± 0.8	1.7 - 9.3
C57BL/6J:BX	SUSCEPTIBLE	0.03 ± 0.06 (6)	4.81 ± 1.24 (11)	2.31 - 11.93	0.4 ± 0.2	19.8 ± 1.9	14.0 - 31.7
	RESISTANT		1.46 ± 0.21 (11)	0.66 - 1.90		6.6 ± 0.7	3.3 - 9.7
B6C3F2	SUSCEPTIBLE	0.02 ± 0.02 (8)	4.80 ± 1.00 (15)	2.49 - 14.29	0.5 ± 0.2	25.1 ± 2.5	13.3 - 47.3
	RESISTANT		0.68 ± 0.09 (42)	0.03 - 1.83		4.7 ± 0.4	0.3 - 10.3

[a]Phenotypes with a response greater than, or equal to, the B6 progenitor strain were designated SUSCEPTIBLE. Phenotypes with a response less than the B6 progenitor strain were designated RESISTANT.
[b]Numbers in parentheses represent sample sizes.
*Statistical comparisons of C3H/HeJ with C57BL/6J: $P < 0.001$. B6C3F1/J: B6 female × C3 male; C3B6F1: C3 female × B6 male; C3H/HeJ:BX: C3 female × B6C3F1/J male; C57BL/6J:BX; B6 female × B6C3F1/J male; B6C3F2: B6C3F1/J female × B6C3F1/J male.

TIBLE phenotypes, and progeny of the F_2 generation should exhibit a 3:1 ratio. Based on the lower limits of the response by B6 mice, we designated as SUSCEPTIBLE all mice which have > 11% or > 2.25×10^4 PMNs in total cells recovered, and RESISTANT all mice with < 11% or < 2.25×10^4 PMNs. Offspring of the C3 back-cross were thus uniformly RESISTANT and not significantly different from C3 mice (Table 6.1), and progeny of the B6 backcross segregated into RESISTANT and SUSCEPTIBLE phenotypes with a frequency of 11:11 (Table 6.1). These ratios are not significantly different from predicted ratios. Progeny of the F_2 generation segregated into the RESISTANT and SUSCEPTIBLE phenotypes with a frequency of 43:15 (74%:26%; Table 6.1), which is not significantly different from the predicted 3:1 ratio. No sex-related differences in the inflammatory responses have been detected among the parental strains or any of the progeny of crosses between them.

The observed phenotypes of the back-cross and F_2 progeny are, therefore, consistent with the hypothesis that a single autosomal recessive gene (which we have termed *Inf*) may confer susceptibility to acute oxidant-induced lung inflammation. This unique model thus provides the potential to apply breeding techniques and cosegregation (linkage) analysis (see below) to investigate mechanisms of susceptibility to ozone-induced inflammation.

It may be hypothesized that a previously characterized locus which controls inflammatory response(s) to other agents in inbred mouse strains may also control ozone-induced inflammation. For example, the inflammatory response to lipopolysaccharide (LPS) challenge is controlled by a single locus (*Lps*), and B6 and C3 mice have been found to be susceptible and resistant, respectively.

To test the hypothesis that the *Inf* and *Lps* loci are the same, we compared the responses to ozone and LPS in six strains of mice (Fig. 6.2). The response to ozone depicted in Fig. 6.2 represents the maximum detected influx of PMNs following a 3-hour exposure to 2 ppm ozone. The responses to LPS represent a compilation of data from previously published studies which have sought to determine the mechanism of control of the *Lps* locus (Watson *et al.*, 1977; Brandwein *et al.*, 1987; MacVittie and Weinberg, 1980; Vogel *et al.*, 1980). These studies have examined a number of parameters of response to LPS and have found distinct strain-specific phenotypes (SUSCEPTIBLE and RESISTANT) for each parameter. As shown in Fig. 6.2, the strain-specific response to ozone does not correlate closely with the response(s) to LPS. The C3, B6 and BALB/cJ strains exhibit concordant responses to each of the agents, but C3H/HeN, C3HeB/FeJ and A/J mice displayed differential susceptibility. These data thus suggest that the responses to ozone and

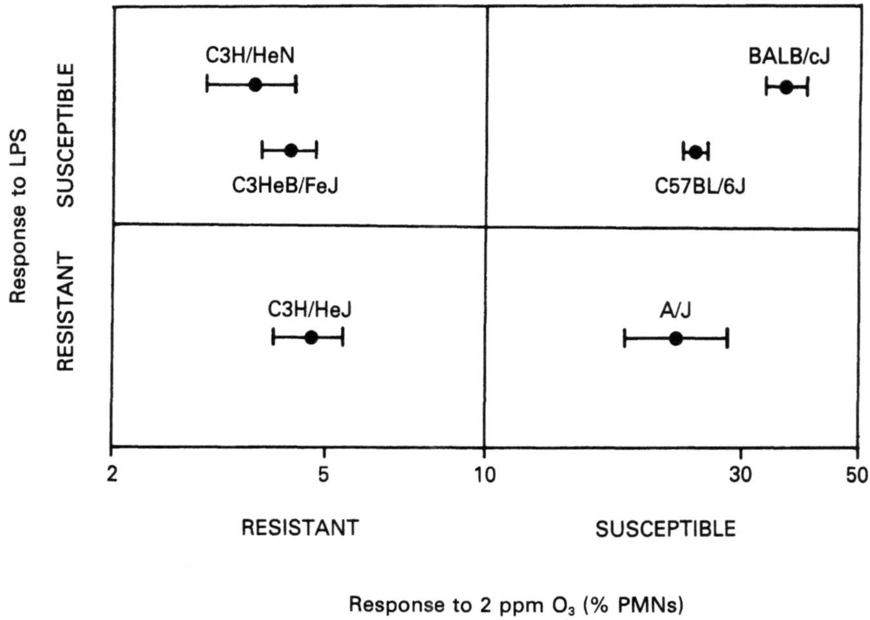

Figure 6.2 Comparisons of responses to lipopolysaccharide challenge and 2 ppm ozone in six highly inbred strains of mice, results expressed as means ±1 SEM. A detailed explanation of the response to each of these agents is given in the text.

LPS are not under the control of the same gene, and the hypothesis that *Inf* and *Lps* are identical may be rejected.

6.4.3 BXH recombinant inbred (RI) analysis

To test further the hypothesis that a single gene controls susceptibility and to initiate linkage analysis, we typed 12 BXH recombinant inbred (RI) strains of mice for the change in PMNs 6 hours after acute ozone, and compared their responses to progenitor C3 and B6 mice (Kleeberger *et al.*, 1989). The ozone-induced PMN response of 8 of 12 RI strains did not differ significantly (99% confidence) from progenitor C3 mice and were typed as RESISTANT (Fig. 6.3). three RI strains had PMN responses which were not significantly different from the B6 progenitor strain and were typed as SUSCEPTIBLE. The response of one RI strain was significantly greater than the B6 strain and may be typed HYPERSUSCEPTIBLE. This strain distribution pattern (SDP) suggests the possibility of another locus which may regulate the degree of susceptibility to ozone-induced inflammation. Comparison of the *Inf* SDP with other BXH RI SDPs indicated significant concordance of the

Phenotypes

C57BL/6J ■
C3H/HeJ □

Locus	*Chromosome*	2	3	4	6	7	8	9	10	11	12	14	19
		BXH recombinant inbred strains											
Ncam	09	■	■	□	■	□	■	□	□	□	□	■	□
MCR	09	□	□	□	■	□	■	□	□	■	□	■	□
Inf	09	□	□	□	■	□	■	□	□	■	□	■	□
MOD–1	09	□	□	□	□	□	■	□	□	■	□	■	□

Figure 6.3 The BXH RI strain distribution pattern (SDP) obtained for four different loci. The closed squares represent phenotypes which are similar to the C57BL/6J progenitor strain, and the open squares represent phenotypes which resemble the C3H/HeJ strain.

Inf locus with three previously described loci on chromosome nine (*MCR, NCAM, Mod–1*; Fig. 6.3), and a tentative map assignment of the *Inf* locus has thus been made to this chromosome. As demonstrated in the present studies, BXH RIs are an extremely valuable tool for linkage analysis and mapping studies.

6.5 HYPOTHESIS TESTING WITH GENETIC MODELS

One of the great strengths of a genetic model is the ability to test hypotheses about mechanisms of the response under study, such that a primary defect which leads to the disorder may be identified. In Mendelian disorders, two research strategies have traditionally been used to determine the primary defect which results in the pathophysiological process. The classical approach ('top down' or 'forward genetics') is to identify a biochemical abnormality which may result from an enzyme deficiency. Ultimately, the gene for the enzyme is identified and cloned, and the change in the gene determined. The globin genes and the gene for alpha$_1$-antitrypsin have been identified and characterized following this approach. Another approach ('bottom up' or 'reverse genetics') is to go directly to the gene by mapping studies, determine the structure and function of the normal counterpart of the gene that is mutant in the disorder, and then work back to the expressed phenotype (McKusick, 1988). The discovery of the gene for cystic fibrosis is a recent example of the powerful utility of this approach (Buchwald *et al.*, 1989; Rommens *et al.*, 1989).

Single-gene models are particularly useful because the techniques of cosegregation and candidate gene analyses may be utilized to determine mechanisms of gene action and to determine chromosomal location of the gene (reverse genetics). Cosegregation analysis has been used to develop an understanding of the molecular mechanisms of such diverse phenomena as toxic chemical depression of the bone marrow (Nebert *et al.*, 1977), variation in metabolism of chemical carcinogens (Levitt *et al.*, 1979) and tumour susceptibility (Nebert *et al.*, 1979).

Candidate gene analysis has also been used extensively to investigate potential mechanisms of disease processes, and details of the application and use of these analyses are very well described by McKusick (1988). One of the strategies of this approach is to determine whether the disorder in question maps to the region of a candidate gene, i.e. a previously characterized gene which has apparent relevance to the disorder. If the locus maps to the region of the candidate gene, a potential mechanism for the disorder may be identified. If a match is not detected, for example if the two loci are on different chromosomes, then the investigator may exclude the candidate gene as a potential pathophysiological mechanism for the disorder in question. The investigator thus economizes effort and helps direct research toward more tenable hypotheses.

The usefulness of this approach may be illustrated with our ozone susceptibility model. Because the *Inf* locus has been found to have significant linkage with loci on mouse chromosome 9, and because the gene which controls LPS responsiveness in the mouse (*Lps*) is located on chromosome 4, the hypothesis that the two loci are the same may be rejected (the same conclusion reached by strain comparison, as described above). After the map assignment of a Mendelian locus is verified, the candidate gene approach thus provides a very powerful tool for testing hyotheses about the mechanism of control of the locus, as with the *Inf*.

6.6 CONCLUSIONS

The utility of animal models of genetic susceptibility to environmental chemicals has been recognized since biological variability in responsiveness to these agents was noted. Surprisingly, few animal models have been well-characterized genetically, and the full potential of these models for predicting genetically-determined hypersusceptibility in the general population has not yet been realized. However, as discussed above, animal models provide an excellent means to investigate the mechanisms of genetic predisposition to the effects of chemical exposures and to determine methods of intervention or inhibition of the subsequent consequences of exposures.

Single-gene, or Mendelian, murine models are particularly useful because there is approximately 80% homology between human and mouse genomes. A primary objective of these models may be to establish linkage associations of genetically-determined susceptibility factors with phenotypic markers which are easily obtainable and analogous in the mouse and human. Such markers may help to identify individuals who are at risk to oxidant exposures.

Animal models of genetic susceptibility therefore have very important implications for public health, and great effort should be made to develop and characterize them. These efforts should aim to link the predictive capabilities of animal models with human studies. Investigation of the mechanisms which produce the analogous phenotype(s) in the animal model may provide significant insight into human pathophysiological processes.

ACKNOWLEDGEMENTS

Dr Kleeberger's research was supported by NIH Grants ES–03505, ES–03819 and R23 HL–37061. Dr Levitt was supported by the Merck Foundation.

REFERENCES

Bates, D. V. (1972) Air pollutants and the human lung. *Am. Rev. Respir. Dis.*, **105**, 1–13.

Brandwein, S. R., Skamene, E., Aubut, J. *et al.* (1987) Genetic regulation of lipopolysaccharide-induced interleukin–1 production by murine peritoneal macrophages. *J. Immunol.* **138**, 4263–9.

Brieger, H. (1963) Genetic bases of susceptibility and resistance to toxic agents. *J. Occup. Med.*, **5**, 511–15.

Briles, D. E., Benjamin Jr, W. H., Huster, W. H. and Posey, B. (1986) Genetic approaches to the study of disease resistance: with special emphasis on the use of recombinant inbred mice. *Curr. Top. Microbiol. Immunol.*, **124**, 21–35.

Brown, S. L., Brett, S. M., Gough, M. *et al.* (1988) Review of interspecies risk comparisons. *Regul. Toxicol. Pharmacol.*, **8**, 191–206.

Buchwald, M., Tsui L-C. and Riordan, J. R. (1989) The search for the cystic fibrosis gene. *Am. J. Physiol.*, **257**, L47-L52.

Calabrese, E. F. (1978) *Pollutants and High Risk Groups*, Wiley, New York.

Calabrese, E. F. (1984) *Ecogenetics: Genetic Variation in Susceptibility to Environmental Agents*, Wiley, New York.

Cheers, C., McKenzie, I. F. C., Mandel, T. E. and Chan, Y. Y. (1980) A single gene (Lr) controlling natural resistance to murine listeriosis, in *Genetic Control of Natural Resistance to Infection and Malignancy*. (eds E. Skamene, P. A. L. Kongshavn and M. Landy), Academic Press, New York, pp. 141–7.

Clark, D. W. J. (1985) Genetically determined variability in acetylation and oxidation. Therapeutic implication. *Drugs*, **29**, 342–75.

Clayson, D. B. (1985) Problems in interspecies extrapolation, in *Toxicological Risk Assessment. Vol I. Biological and Statistical Criteria* (eds D. B. Clayson, D. Krewski and I. Munro), CRC Press, Boca Raton FL, 105–22.

Costa. D. L., Schafrank, S. N., Wehner, R. W. and Jellett, E. (1985) Alveolar permeability to protein in rats differentially susceptible to ozone. *J. Appl. Toxicol.*, **5**, 182–6.

Crouch, E. A. C. (1983a) Carcinogenic risk assessment – the consequences of believing models. *Basic Life Sci.*, **24**, 653–65.

Crouch, E. A. C. (1983b) Uncertainties in interspecies extrapolations of carcinogenicity. *Environ. Health Perspect.*, **50**, 321–7.

Festing, M. F. W. (1979) Properties of inbred strains and outbred stocks, with special reference to toxicity testing. *J. Toxicol. Environ. Health*, **5**, 53–68.

Festing, M. F. W. (1987) Genetic factors in toxicology: implication for toxicological screening. *CRC Crit. Rev. Toxicol.*, **18**, 1–26.

Folinsbee, L. J., Bedi, J. F. and Horvath, S. M. (1984) Pulmonary function changes after 1 h continuous heavy exercise in 0.21 ppm ozone. *J. Appl. Physiol.*, **57**, 984–8.

Fox, M. R. S. (1979) Nutritional influences on metal toxicity: cadmium as a model toxic element. *Environ. Health Perspect.*, **29**, 95–104.

Gardner, D. E. (1979) Introductory remarks: session on genetic factors affecting pollutant toxicity. *Environ. Health Perspect.*, **29**, 45–8.

Gervais, F., Stevenson, M. M. and Skamene, E. (1984) Genetic control of resistance to *Listeria monocytogenes*: regulation of leukocyte inflammatory response by the *Hc* locus. *J. Immunol.*, **132**, 2078–83.

Goldstein, B. D., Lai, L. Y., Ross, S. R. and Cuzzi-Spada, R. (1973) Susceptibility of inbred mouse strains to ozone. *Arch. Environ. Health*, **27**, 412–13.

Gould, M. N. (1986) Inheritance and site of expression of genes controlling susceptibility to mammary cancer in an inbred rat model. *Cancer Res.*, **46**, 1199–1202.

Greaves, J. F. and Ayres, P. B. (1977) Unifactorial inheritance of warfarin resistance in *Rattus norvegicus* from Denmark. *Genet. Res.*, **29**, 215–22.

Green, M. C. (1981) *Genetic Variants and Strains of the Laboratory Mouse*, Gustav Fischer Verlag, Stuttgart.

Gregory, A. R. (1988) Species comparisons in evaluating carcinogenicity in humans. *Regul. Toxicol. Pharmacol.*, **8**, 160–90.

Hansen, M. F. and Cavenee, W. K. (1987) Genetics of cancer predisposition. *Cancer Res.*, **47**, 5518–27.

Hargreave, F. E., Ramsdale, E. H., Kirby, J. G. and O'Byrne, P. M. (1986) Asthma and the role of inflammation. *Eur. J. Respir. Dis.*, **69** (Suppl. 147), 16–21.

Hart, R. W. and Fishbein, L. (1985) Interspecies extrapolation of drug and genetic toxicity data, in *Toxicological Risk Assessment. Vol I. Biological and Statistical Criteria* (eds D. B. Clayson, D. Krewski and I. Munro), CRC Press, Boca Raton FL, pp. 3–40.

Hein, D. W., Omichiniski, J. G., Brewer, J. A. and Weber, W. W. (1982) A unique pharmacogenetic expression of the *N*-acetylation polymorphism in the inbred hamster. *J. Pharmacol. Exp. Ther.*, **220**, 8–15.

Henningsen, N. C., Cederberg, A., Hanson, A. and Johannson, B. W. (1975) Effects of long term treatment with procaine amide. A prospective study with special regard to ANF and SLE in fast and slow acetylators. *Act Med. Scand.*, **198**, 475–82.

Hogan, M. D. (1983) Extrapolation of animal carcinogenicity data: limitations and pitfalls. *Environ. Health Perspect.*, **47**, 333–7.

Idle, J. R. and Smith, R. L. (1980) Inter-ethnic differences in the metabolic disposition of drugs and toxic substances, in *Toxicology in the Tropics* (eds R. L. Smith and E. A. Babaunmi), Taylor and Francis, London, pp. 239–53.

Kalow, W. (1962) *Pharmacogenetics: Heredity and the Response to Drugs*, Saunders, Philadelphia.

Kellerman, G., Luyten-Kellerman, M. and Shaw, C. R. (1973a) Genetic variation of aryl hydrocarbon hydroxylase in human lymphocytes. *Am. J. Hum. Genet.*, **25**, 327–31.

Kellerman, G., Shaw, C. R. and Luyten-Kellerman, M. (1973b) Aryl hydrocarbon hydroxylase inducibility and bronchogenic carcinoma. *N. Eng. J. Med.*, **289**, 934–7.

Kleeberger, S. R., Bassett, D. J. P., Jakab, G. J. and Levitt, R. C. (1990) A genetic model for the evaluation of susceptibility to ozone-induced inflammation. *Am. J. Physiol.*, **258**, 313–20.

Kleeberger, S. R., Hudak, B. B. and Levitt, R. C. (1989) Genetic control of susceptibility to acute ozone exposure: regulation of the polymorphonuclear leukocyte (PMN) inflammatory response by the *Inf* locus. *Federation Amer. Soc. Exp. Biol. J.*, **3**, A1147.

Kleeberger, S. R., Kolbe, J., Adkinson Jr, N. F. *et al.* (1988) The role of mediators in the response of the canine peripheral lung to 1 ppm ozone. *Am. Rev. Respir. Dis.*, **137**, 321–5.

Koenig, J. Q. and Omenn, G. S. (1988) Genetic factors, in *Variations in Susceptibility to Inhaled Pollutants* (eds J. D. Brain, B. D. Beck, A. J. Warren and R. A. Shaikh), Johns Hopkins University, Baltimore, pp. 59–88.

Levitt, R. C., Pelkonen, O., Okey, A. B. and Nebert, D. W. (1979) Genetic differences in metabolism of polycyclic aromatic carcinogens and aromatic amines by mouse liver microsomes. Detection by DNA binding of metabolites and by mutagenicity in histadine-dependent *Salmonella typhimurium* in vitro. *J. Natl. Cancer Inst.*, **62**, 947–55.

Lush, I. E. (1981) Mouse pharmacogenetics. *Symp. Zool. Soc. Lond.*, **47**, 517–46.

McDonnell III, W. F., Horstman, D. H., Abdul-Salaam, S. and House, D. E. (1985) Reproducibility of individual responses to ozone exposure. *Am. Rev. Respir. Dis.*, **131**, 36–40.

McKusick, V. A. (1983) *Mendelian Inheritance in Man*, Johns Hopkins University, Baltimore.

McKusick, V. A. (1988) The morbid anatomy of the human genome: a review of gene mapping in clinical medicine. *Medicine*, **67**, 1–19.

MacVittie, T. J. and Weinberg, S. R. (1980) An LPS responsive cell in C3H/HeJ mice: the peritoneal exudate-derived macrophage colony-forming cell (M-CFC), in *Genetic Control of Natural Resistance to Infection and Malignancy* (eds E. Skamene, P. A. L. Kongshavn and M. Landy), Academic Press, New York, pp. 511–18.

Meier, H. (1963) *Experimental Pharmacogenetics*, Academic Press, New York.

Melmon, K. L. and Bourne, H. R. (1974) Mechanisms of inflammation. *Clin. Pharmacol. Ther.*, **16**, 886–91.

Menzel, D. B. (1979) From animals to man, the grand extrapolation of environmental toxicology, in *Assessing Toxic Effects of Environmental Pollutants* (eds S. D. Lee and J. B. Mudd), Ann Arbor Science, Ann Arbor, pp. 1–14.
Menzel, D. B. (1984) Ozone: an overview of its toxicity in man and animals. *J. Toxicol. Environ. Health*, **13**, 183–204.
Mohn, G. R. (1978) An overview of animal and microbial test systems for carcinogenesis and mutagenesis. Problems with human variation. *Hum. Genet. (Suppl.)*, **1**, 169–76.
Mortensen, R. F., Le, P. T. and Taylor, B. A. (1985) Mouse serum amyloid P-component (SAP) levels controlled by a locus on chromosome 1. *Immunogenetics*, **22**, 367–75.
Motulsky, A. G. (1977) Ecogenetics: genetic variation in susceptibility to environmental agents, in *Human Genetics* (eds S. Armandares and R. Lisker), Excerpta Medica, Amsterdam, pp. 375–85.
Motulsky, A. G. (1978) Bioethical problems in pharmacogenetics and ecogenetics. *Hum. Genet. (Suppl.)*, **1**, 185–92.
Mustafa, M. G. and Tierney, D. F. (1978) Biochemical and metabolic changes in the lung with oxygen, ozone, and nitrogen dioxide toxicity. *Am. Rev. Respir. Dis.*, **118**, 1061–90.
Nebert, D. W. (1980) Pharmocogenetics. An approach to understanding chemical and biologic aspects of cancer. *J. Natl. Cancer Inst.*, **64**, 1279–90.
Nebert, D. W. and Atlas, S. A. (1978) The *Ah* locus: aromatic hydrocarbon responsiveness . . . of mice and men. *Hum. Genet. (Suppl.)*, **1**, 149–60.
Nebert, D. W., Atlas, S. A., Guenthner, T. M. and Kouri, R. E. (1978) The *Ah* locus: genetic regulation of the enzymes which metabolize polycyclic hydrocarbons and the risk for cancer, in *Polycyclic Hydrocarbons and Cancer: Chemistry, Molecular Biology and Environment* (eds P. O. P. Ts'o and H. V. Gelbion), Academic Press, New York, pp. 345–90.
Nebert, D. W. and Gielen, J. E. (1972) Genetic regulation of aryl-hydrocarbon hydroxylase induction in the mouse. *Fed. Proc.*, **31**, 1315–25.
Nebert, D. W., Levitt, R. C., Jenson, N. M. *et al.* (1977) Birth defects and aplastic anemia: differences in polycyclic hydrocarbon toxicity associated with the *Ah* locus. *Arch. Toxicol.*, **39**, 109–32.
Nebert, D. W., Levitt, R. C. and Pelkonen, O. (1979) Genetic variation in metabolism of chemical carcinogens associated with susceptibility to tumorigenesis, in *Carcinogens: Identification and Mechanisms of Action* (eds A. C. Griffin and C. R. Shaw), Raven Press, New York, pp. 157–85.
NIH (National Institutes of Health) (1985) Mechanisms of inflammation in asthma, *NIH Publ. No. 86-2666.*
Nordberg, G. F. and Kjellstrom, T. (1979) Metabolic model for cadmium in man. *Environ. Health Perspect.*, **28**, 211–17.
O'Brien, A. D., Rosenstreich, D. L. and Taylor, B. A. (1980) Control of natural resistance to *Salmonella typhimurium* and *Leishmania donovani* in mice by closely linked but distinct genetic loci. *Nature*, **287**, 440–2.
Omenn, G. S. (1978) Ecogenetics: a view from the U.S. President's science advisory office. *Hum. Genet. (Suppl.)*, **1**, 179–84.
Omenn, G. S. and Motulsky, A. G. (1978) Ecogenetics: genetic variation in the susceptibility to environmental agents, in *Genetic Issues in Public Health and Medicine* (eds B. H. Cohen, A. M. Lilienfeld and P. C. Huang), C. C. Thomas, Springfield, pp. 83–111.
Pauli, R. M., Lian, J. B., Moshier, D. F. and Suttie, J. W. (1987) Association of congenital deficiency of multiple vitamin K-dependent coagulation factors and

the phenotype of the warfarin embryopathy: clues to the mechanism of teratogenicity of coumarin derivatives. *Am. J. Hum. Genet.*, **41**, 566–83.

Perry, Jr, H. M., Tan, E. M., Carmody, S. and Sakamoto, H. (1970) Relationship of acetyl transferase activity to antinuclear antibodies and toxic symptoms in hypertensive patients treated with hydralazine. *J. Lab. Clin. Med.*, **76**, 114–24.

Plant, J. E., Blackwell, J. M., O'Brien, A. D. *et al.* (1982) Are the *Lsh* and *Ity* disease resistance genes at one locus on mouse chromosome 1? *Nature*, **297**, 510–11.

Poland, A. P., Glover, E., Robinson, J. R. and Nebert, D. W. (1974) Genetic expression of aryl-hydrocarbon hydroxylase activity. *J. Biol. Chem.*, **249**, 5599–606.

Poland, A. P., Glover, E. and Taylor, B. A. (1987) The murine *Ah* locus: a new allele and mapping to chromosome 12. *Mol. Pharmacol.*, **32**, 471–8.

Price-Evans, D. A. (1980) Genetic factors in adverse reactions to drugs and chemicals, in *PAR, Pseudoallergic Reactions, Involvement of Drugs and Chemicals, vol. 1* (eds P. Dukor, P. Kallos, H. D. Schlumberger and G. B. West), S. Karger, Basel, pp. 1–27.

Rommens, J. M., Iannuzzi, M. C., Kerem, B. *et al.* (1989) Identification of the cystic fibrosis gene: chromosome walking and jumping. *Science*, **245**, 1059–65.

Seltzer, J., Bigby, B. G., Stulbarg, M. *et al.* (1986) O_3-induced change in bronchial reactivity to methacholine and airway inflammation. *J. Appl. Physiol.*, **60**, 1321–6.

Shaikh, R. A., Warren, A. J. and Little, J. B. (1988) Genetic factors, in *Variations in Susceptibility to Inhaled Pollutants* (eds J. D. Brain, B. D. Beck, A. J. Warren and R. A. Shaikh), Johns Hopkins University, Baltimore, pp. 223–52.

Skamene, E. (1986) Genetic control of resistance to mycobacterial infection. *Curr. Top. Microbiol. Immunol.*, **124**, 49–66.

Skamene, E., Gros, P., Forget, A. *et al.* (1982) Genetic regulation of resistance to intracellular pathogens. *Nature*, **297**, 506–9.

Tannen, R. H. and Weber, W. W. (1980) Inheritance of acetylator phenotype in mice. *J. Pharmacol. Exp. Ther.*, **213**, 480–4.

Taylor, B. A. (1976) Linkage of the cadmium resistance locus to loci on mouse chromosome 12. *J. Hered.*, **67**, 389–90.

Taylor, B. A. (1980) Recombinant inbred strains of mice: use in genetic analysis of disease resistance, in *Genetic Control of Natural Resistance to Infection and Malignancy* (eds E. Skamene, P. A. L. Kongshavn and M. Landy), Academic Press, New York, pp. 1–7.

Taylor, B. A., Heiniger, H. J. and Meier, H. (1973) Genetic analysis of resistance to cadmium induced testicular damage in mice. *Proc. Soc. Exp. Biol. Med.*, **143**, 629–33.

Thomas, P. E., Louri, R. E. and Hutton, J. J. (1972) The genetics of aryl-hydrocarbon hydroxylase induction in mice: a single gene difference between C57BL/6J and DBA/2J. *Biochem. Genet.*, **6**, 157–68.

Vane, J. and Botting, R. (1987) Inflammation and the mechanisms of action of anti-inflammatory drugs. *FASEB J.*, **1**, 89–96.

Vogel, S. N., Weedon, L. L., Oppenheim, J. J. and Rosenstreich, D. L. (1980) Defective Fc-mediated phagocytosis by LPS-hyporesponsive (Lps^d) C3H/HeJ macrophages: correction by agents that elevate intracellular cyclic AMP, in *Genetic Control of Natural Resistance to Infection and Malignancy* (eds E. Skamene, P. A. L. Kongshavn and M. Landy), Academic Press, New York, pp. 583–9.

Wallace, M. E. and MacSwiney, F. J. (1976) A major gene controlling warfarin-resistance in the house mouse. *J. Hyg. (Camb.)*, **76**, 173–81.

Watson, J., Riblet, R. and Taylor, B. A. (1977) The response of recombinant inbred strains of mice to bacterial lipopolysaccharides. *J. Immunol.*, **118**, 2088–93.

Weber, W. W. (1984) Acetylation pharmacogenetics: experimental models for human toxicity. *Fed. Proc.*, **43**, 2332–7.

Wilson, R. and Crouch, E. A. C. (1984) Problems in interspecies comparisons, in *Mechanisms of DNA Damage and Repair* (eds M. G. Simic, L. Grossman and A. C. Upton), Plenum Press, New York, pp. 543–9.

Woosley, R. L., Drayer, D. N., Reidenberg, M. M. *et al.* (1978) Effect of acetylator phenotype on the rate at which procainamide induces antinuclear antibodies and the lupus syndrome. *N. Engl. J. Med.*, **298**, 1157–9.

7 Sex-related differences in genetic susceptibility to toxic metals

K. Kostial, M. Blanuša, B. Kargačin, M. Piasek, T. Maljković
Institute for Medical Research and Occupational Health, University of Zagreb, Zagreb, Croatia, Yugoslavia

and D. Kello
Environment and Health Service, WHO Regional Office for Europe, Copenhagen, Denmark

7.1 SUMMARY

This chapter summarizes and evaluates the results of a series of studies performed in one laboratory in regard to sex-related differences in toxicokinetics and toxicity of some metals. Experiments were performed on adult male and female albino rats, and on their litters. The toxicokinetics of cadmium, mercury and manganese were studied after single intraperitoneal or oral administration of radioisotopes ^{115m}Cd, ^{203}Hg and ^{54}Mn. The effect of lactation on gastrointestinal absorption of lead and strontium was studied after single oral administration of radioisotopes ^{203}Pb and ^{85}Sr. The sex-related difference in lead toxicity was studied after oral exposure to 1500 ppm, 3500 ppm, 5500 ppm, or 7500 ppm of lead in drinking water. The results show that toxicokinetics and toxicity of metals studied are sex-dependent. These data strongly suggest that sex-related differences in susceptibility to toxic substances must be considered in setting environmental or occupational standards.

7.2 INTRODUCTION

All international recommendations related to the limit values for toxic chemicals in the environment should, by definition, protect all individuals, including those who are highly susceptible (WHO, 1984). In reality, however, current methods of risk assessment are based on the extrapolations of 'response data' of the average individual, ignoring completely the phenotypic variations and, therefore, particularly susceptible subgroups of the population.

Different methods are used in risk assessment procedures in an attempt to overcome this problem and to protect all segments of the population. In the absence of a better method, 'safety factors' are traditionally used to cover a variety of uncertainties, including possible undetected effects on particularly sensitive members of the population. Traditionally, safety factors were used to allow for uncertainties in extrapolation from laboratory animals to humans, and from a small group of individuals to a large population (WHO, 1987).

Differences in response can be attributed to many factors, such as pre-existing health or physiological conditions, other confounding factors and multiple exposures, but above all to the unknown factors, such as individual predisposition. A population must be considered very heterogeneous in respect to response to pollutants, but existing scientific information does not often allow adequate assessment of the proportion of the population that has enhanced response. However, an estimate of even a few per cent of the total population would translate to a large number of people (WHO, 1987).

Although substantial theoretical foundations support the hypothesis that genetic conditions are likely to enhance the risk of environmentally- or occupationally-induced adverse health effects, experimental and epidemiological evidence is still insufficient for appropriate regulatory actions. Of the approximately 50 human genetic conditions identified as potentially responsible for the genetic susceptibility, only G–6-PD deficiency has a demonstrable causal history of enhancing individual susceptibility to some industrial pollutants (see Chapter 2 and Calabrese, 1984).

Between 10 and 20% of the normal young adult population may be 'hyperreactors' (Calabrese, 1984). However, if this hypersensitivity is linked to a specific sex difference, than the segment of population at risk increases to 50%. Therefore, gender-related genetic predisposition to toxic chemicals has special importance in the prevention and control of environmentally- or occupationally-induced health hazards. Sex-related differences in response to toxic substances have been described in a sizeable number of species and for a large number of toxic substances. Given the substantial influx of women into traditionally male-dominated occupations, the assessment of sex-related differences in response to toxic substances becomes exceptionally important (Calabrese, 1985).

Although sex-related differences in response to toxic substances have been suspected since the early 1900s, the first comprehensive scientific review on this topic was by Calabrese in 1985. The author undertook a critical assessment of the literature on biomedical and physiological processes in pregnancy, liver and kidney toxicity and reactions to inorganic and organic contaminants, drugs, oral contraceptives and some other endogenous substances, with particular emphasis on sex-related differences. According to this review, the number of toxic substances for which sex-related differences in susceptibility are described in the literature approaches 200, and such differences are found in many species. Krasowskii analysed the sex-related differences in the response of white rats exposed to 149 chemicals (Krasowskii, 1976). For the majority of substances there was a two- to three-fold difference in response, however with certain organophosphorus compounds the females were up to four times more sensitive than males.

Considerable progress has been made in the elucidation of mechanisms in sex-specific responses. The causes of these responses have been related to differences in a variety of physiological processes, in enzyme activities as well as in genetic repair mechanisms. The sex-linked differences in respect to renal toxicity have been studied most extensively. Of the 11 toxins given, the male rat was more sensitive to ten and the female to one (Calabrese, 1985). A comparable phenomenon exists in respect to liver toxicity. Calabrese (1985) compiled six examples of sex-specific differences in xenobiotic-induced liver toxicity, indicating that in five out of six cases the female is more susceptible. This analysis suggests that males are at greater risk to renal toxins and females to liver toxins.

This chapter presents the results from a series of experiments on rats of sex-related differences in toxicokinetics and toxicity of some metals. In the light of these findings, as well as on the bases of other scientific evidence, specific proposals for changes in the current risk assessment practice are discussed.

7.3 METHODS

All experiments were performed in young or sexually mature male and female albino Wistar derived rats, bred in the Institute's breeding farm. Animals were kept in macrolon cages provided with stainless steel covers and floors (Ehret, Vienna, Austria) and with free access to drinking water and standard pelleted food ('Sljeme', Zagreb, Yugoslavia) and in some experiments in individual metabolic cages.

Sex-related differences in toxicokinetics of cadmium, mercury and manganese

Mature rats, between 4 and 5 months old were used in this series of experiments. Each animal was placed in a separate metabolic cage to avoid external contamination. All rats received radioactive isotopes of different metals by a single intraperitoneal injection and were sacrificed 9–14 days later. The radioisotopes ^{115m}Cd, ^{203}Hg and ^{54}Mn were administered as chloride solutions in a volume of about 0.5 ml. The whole body radioactivity was determined in a double crystal scintillation counter (Tobor, Nuclear Chicago) immediately after radioisotope administration and 9–14 days later. Results were expressed as a percentage of the initial whole body retention. At the end of the experiment the animals were sacrificed and the total hair, including the skin, was removed, both kidneys and the liver were dissected and the radioactivity was determined. Results were presented as a percentage of the dose and a percentage of the whole body radioactivity. All results were corrected for radioactive decay and geometry of the sample.

Sex-related differences in reproductive and haemopoietic responses to lead

In these experiments lead acetate was administered in demineralized drinking water over a period of up to 20 weeks. Exposure levels were 1500–7500 ppm lead. The controls received demineralized drinking water. The animals were bred 4 weeks after the beginning of exposure to lead. The control and exposed females were mated with the control and exposed males.

The reproductive parameters determined were incidence of pregnancy (%), number of live newborn and 11-day-old pups per litter and pup body weight (g). The other health-related parameters were food and water consumption, body weights of adult animals and haemopoietic response to lead. Haemopoietic response was studied in rats exposed only to 1500 ppm lead acetate in drinking water for a 4-week and a longer 14-week interval. At the end of both exposures blood was sampled from at least nine rats in each group by incision of the tail vein.

The haematological parameters determined were red blood cell count (RBC), haematocrit (Htc) and haemoglobin (Hb). The concentration of lead in blood (PbB) was estimated each time in other males and females from the same exposure groups. Blood samples were obtained by puncture of the orbital venous plexus behind the eyeball.

The effect of lactation on absorption of lead and strontium

Lactating and control female rats were used in those experiments. During the second week of lactation a single oral dose of ^{203}Pb or ^{85}Sr was given to both groups of animals and they were transferred to individual metabolic cages. Radioactivity was determined using a whole-body, twin-crystal scintillation counter (Tobor, Nuclear Chicago). The recovery of each radioactive dose measured in the whole body and excreta varied from 90 to 96%. The results of apparent absorption of Pb and Sr are expressed as the combined whole body content of mother and pups and cumulative urinary excretion of the respective radionuclides.

The methods used in these experiments are elaborated in more details in the references cited below.

7.4 RESULTS

Effects of gender on the toxicokinetics of ^{115m}Cd, ^{203}Hg and ^{54}Mn

Gender was found to influence the use of hair as an indicator of the body burden of metals (Table 7.1). After a single intraperitoneal administration of ^{115m}Cd or ^{203}Hg, a higher fraction of the administered dose was found in the hair and kidney of males as compared to females (Kostial *et al.*, 1985). Both dietary calcium levels and gender were found to influence the absorption of cadmium. The retention of ^{115m}Cd after a single oral dose was inversely correlated with the dietary calcium intake, but was always found to be lower in males than in females irrespective of the dietary calcium level (Kello *et al.*, 1979).

After a single intraperitoneal administration of ^{54}Mn, elimination of manganese from the whole body was at all intervals much higher in females than in males (Rabar, 1976).

The influence of gender on the haemopoietic and reproductive responses to lead

Rats orally exposed to lead showed sex-related differences in the haemotoxic action of lead (Table 7.2). At the exposure level of 1500 ppm in drinking water, reduction of Hb concentration occurred earlier and was more pronounced in female than in male rats (Piasek *et al.*, 1989).

Reproductive studies were performed on male and female rats exposed to lead in drinking water. In females exposed to lead levels above 3500 ppm, the number, weight and survival of pups were reduced compared

Table 7.1 Sex-related differences in toxicokinetics of ^{115m}Cd, ^{203}Hg and ^{54}Mn in rats

Parameter recorded	*Radioisotope*	*Females*[a]	*Males*[a]	*Comment*
Retention 9 days after single ip administration (% dose)				
Hair[b]	^{115m}Cd	3.96 ± 0.10 (9)	5.10 ± 0.24 (12)*	Higher retention of Cd and Hg in hair and kidney in males than in females
Kidney		4.07 ± 0.11 (9)	5.26 ± 0.14 (12)*	
Hair[b]	^{203}Hg	4.37 ± 0.17 (12)	7.34 ± 0.36 (11)*	
Kidney		37.17 ± 0.74 (12)	42.24 ± 0.92 (11)*	
Retention in whole body 14 days after single po administration (% dose)[c]				
Low Ca diet (0.3%)	^{115m}Cd	0.48 ± 0.09 (8)	0.25 ± 0.15 (18)	Higher gastrointestinal absorption of Cd in females than in males
Normal Ca diet (0.7%)		0.36 ± 0.07 (8)	0.18 ± 0.10 (8)	
Retention 17 days after single ip administration (% dose)[d]	^{54}Mn	14.8 ± 1.1 (10)	25.20 ± 0.30 (10)*	Higher elimination of Mn in females than in males

[a]Results are presented as arithemetic means ± SEM (number of animals in brackets); age of animals about 4 months.
[b]From Kostial *et al.*, 1985.
[c]From Kello *et al.*, 1979.
[d]From Rabar, 1976.
*Statistical difference between females and males (Student's t-test): $P < 0.001$.

Table 7.2 Sex-related differences in haemopoietic and reproductive responses to lead in rats

Parameter recorded	*Exposure conditions*	*Female*[a]	*Males*[a]	*Comment*
Haemoglobin (%)[b]	Controls	15.0 ± 0.1 (112)	14.9 ± 0.1 (65)	Higher haemopoietic response in females
	1500 ppm Pb orally	13.8 ± 0.2 (10)*	14.6 ± 0.2 (9)	
Pups 11-day-old[c]				
No of live pups per litter	Controls	8.4 ± 0.3 (119)	8.4 ± 0.3 (119)	
Weight of pups (g)		17.9 ± 0.3 (115)	17.9 ± 0.3 (115)	
Survival rate 1–11 days after birth (%)		92.9 ± 1.8 (115)	92.9 ± 1.8 (115)	
No. of live pups per litter	7500 ppm Pb orally	6.0 ± 0.4 (62)*	8.0 ± 0.4 (31)	Higher reproductive response in females
Weight of pups (g)		15.3 ± 0.4 (53)*	18.2 ± 0.4 (31)	
Survival rate 1–11 days after birth (%)		71.1 ± 5.2 (53)*	98.9 ± 0.6 (31)	

[a]Results are presented as arithmetic means ± SEM (in brackets number of animals in haemoglobin values and number of litters in reproductive parameters); age of adult animals about 4–5 months.
[b]From Piasek *et al.*, 1989.
[c]From Šarić *et al.*, 1987; Piasek and Kostial, 1987.
*Identified as statistically different from respective control value (Student's t-test): $P < 0.001$.

to controls (Šarić *et al.*, 1987). In males, no effect on reproduction was observed even at the highest exposure levels, i.e. 5500 and 7500 ppm (Piasek and Kostial, 1987; Šarić *et al.* 1987).

Effect of lactation on absorption of ^{203}Pb and ^{85}Sr from the gastrointestinal tract

Oral administration of ^{203}Pb or ^{85}Sr to lactating females resulted in a two- to three-fold higher absorption from the gastrointestinal tract than in control animals (Table 7.3). Since lactating animals consume about twice as much food as controls, the actual intake of these elements absorbed from the diet is probably between four and six times higher in lactating than in non-lactating animals (Kostial *et al.*, 1969; Kostial and Momčilović, 1972).

7.5 DISCUSSION

On the topic of inorganic environmental pollutants, and toxic metals in particular, substantial scientific evidence indicates sex-related differences in toxicokinetics and toxicity. Johns *et al* reported in 1923 that male rats are considerably more sensitive to cadmium-induced lethality than females. Selective toxicity was also observed in studies investigating testicular damage (Parizek, 1956; Gunn *et al.*, 1965). In another series of studies dealing with sex-related differences in cadmium toxicity, the duration of hexobarbital-induced sleeping time was repeatedly prolonged in male rats but not in females after treatment with different doses of cadmium (Hadley *et al.*, 1974; Johnston *et al.*, 1975; Schnell *et al.*, 1976).

Hepatic microsomal metabolism of hexobarbital was significantly diminished in male rats but not in females following cadmium administration. This phenomenon was extensively studied in relation to cytochrome P–450 and aminolevulinic acid dehydratase synthetase (ALA-S), but evidence about these interrelationships was not conclusive (Schnell *et al.*, 1976; Pence *et al.*, 1977). However, these studies offered several possible explanations for why males are more sensitive to cadmium than females and also highlighted specific pharmacokinetic aspects. Of significance is the study that showed that metallothionein, which binds cadmium and is mainly responsible for its distribution in the organism, occurs in low concentrations in the liver of the normal female rat but is absent in the normal male liver (Webb, 1972). These data could also be of importance for interpretation of our results (Table 7.1) indicating a higher level of available cadmium in males than in females.

That the health effect of mercury may be sex-linked was first reported in 1964 by Harber and Jennings, who showed a significant sex-related

Table 7.3 Effect of lactation on absorption of lead and strontium in rats

Parameter recorded	*Radioisotope (% dose)*	*Controls*[a]	*Lactating rats*[a]	*Comment*
Cumulative 5-day excretion and body retention in mother and litter of orally dosed ^{203}Pb in rats[b]	^{203}Pb	0.76 ± 0.16 (9)	2.61 ± 0.69 (9)*	Gastrointenstinal absorption of Pb increased in lactation
Cumulative 2-day excretion and body retention in mother and litter of orally dosed ^{85}Sr in rats[c]	^{85}Sr	4.98 ± 0.36 (41)	10.68 ± 0.60 (27)*	Gastrointestinal absorption of Sr increased in lactation

[a]Results are presented as arithmetic means ± SEM (number of animals in brackets).
[b]From Kostial and Momčilović, 1972.
[c]From Kostial *et al.*, 1969.
*Identified as statistically different from respective control value (Student's t-test): $P < 0.001$.

difference in the renal tubule toxicity in intravenously injected rats. In contrast to females, the male rats displayed severe coagulative necrosis of the proximal tubules. Muraoka and Itok (1980) confirmed these findings and demonstrated that renal levels of sulfhydryl groups are higher in female rats and that castrated male rats have elevated renal sulfhydryl levels. It was hypothesized that the sex difference in the levels of sulfhydryl groups may be one of the factors affecting differential susceptibility to mercury between males and females, as sulfhydryl compounds are known to lower the toxicity of heavy metals including mercury compounds. This finding could be equally applicable to the sex-related differences in cadmium toxicokinetics and toxicity. As for cadmium, the role of metallothionein in the toxicokinetics of mercury has also been extensively studied.

Sex-related differences in lead toxicity have been described in many publications, and it has long been suspected that women are more susceptible than men to lead intoxication (Oliver, 1902). More substantial scientific evidence about sex-related differences in lead toxicity came from the studies on free-erythrocyte protoporphyrin (FEP). Higher FEP levels at lower blood lead (PbB) levels were found in adult women compared to adult men, and a proportional increase of FEP was associated with the PbB increase (Stuik and Zielhuis, 1975). A large number of studies support these findings. The role of this enhanced susceptibility to lead-induced haematological changes in the proposed greater sensitivity of women to lead-induced neurological changes is not yet fully clarified, but it certainly should be of high public health concern (Calabrese, 1985). The enhanced female response to lead-induced changes in haeme biosynthesis has also been demonstrated in dogs (Maxfield *et al.*, 1972) and rabbits (Wibowo and Zielhuis, 1980). A higher reproductive response in females was also observed by other authors (Stowe and Goyer, 1971). Results of our studies (Table 7.2) are in line with previously published investigations, indicating that the haemopoietic and reproductive responses to lead are more pronounced in females than in males.

The above evidence strongly suggests that sex-related differences in susceptibility to toxic substances must be always considered in risk assessment and risk management decision-making. These differences are of considerable public health significance as they range in magnitude from 10 up to 100% (or a 10-fold difference). However, evidence indicates that these differences could be even higher, as in the case of decalin to which the female rat is at least 100 times less sensitive than the male rat in terms of renal toxicity. Based on this evidence and in the absence of actual data on the response to toxic chemicals for both sexes and in several species, the uncertainty factor (safety factor) used in the regulatory process should incorporate at least an additional factor of five on the basis of possible sex-related differences.

Finally, increased awareness of genetic conditions that predispose affected individuals to develop environmentally- or occupationally-induced ill health emphasises the need for improved epidemiological studies or studies on laboratory animals that will take into account possible sex-related differences. The published research on such differences has mainly focused on experiments in rats, which suggests extrapolation of available data may be limited if the principle that a 'finding described in only one species should be considered as a species-specific phenomenon' is strictly followed. Double-sex studies should be required without exception in licensing the production and marketing of toxic chemicals. However this basic principle, followed in most routine toxicity testings, is very often neglected when risk assessors and regulators use scientific data published in the highly regarded journals. Therefore this fact should be considered as well by the scientific community when conducting toxicological experiments.

REFERENCES

Calabrese, E. J. (1984) *Ecogenetics. Genetic Variation in Susceptibility to Environmental Agents*, John Wiley, New York.

Calabrese, E. J. (1985) *Toxic Susceptibility, Male/Female Differences*, John Wiley, New York.

Gunn, S. A., Gould, T. C. and Anderson, W. A. D. (1965) Strain differences in susceptibilities of mice and rats to cadmium-induced testicular damage. *J. Reprod. Fertil*, **10**, 273–5.

Hadley, W. M., Miya, T. S. and Bousquet, W. F. (1974) Cadmium inhibition of hepatic drug metabolism in the rat. *Toxicol, Appl. Pharmacol.*, **29**, 284–91.

Harber, M. H. and Jennings. R. B. (1964) Sex differences in renal toxicity of mercury in rat. *Nature*, **201**, 1235.

Johns, C. O., Finks, A. J. and Alsberg, C. L. (1923) Chronic intoxication by small quantities of cadmium chloride in the rat. *J. Pharmacol. Exp. Ther.*, **21**, 59–64.

Johnston, R. E., Miya, T. S. and Schnell, R. C. (1975) Cadmium potentiation of drug response – role of the liver. *Biochem. Pharmacol.*, **24**, 879–81.

Kello, D., Dekanić, D. and Kostial, K. (1979) Influence of sex and dietary calcium on intestinal cadmium absorption in rats, *Arch. Environ. Health*, **34** 30–3.

Kostial, K., Blanuša, M. and Landeka, M. (1985) The influence of sex on retention of metals in hair, in *Proceedings of the 5th International Conference 'Heavy Metals in the Environment'* (ed. T. D. Lekkas), Vol. 2, Athens, pp. 116–18.

Kostial, K., Gruden, N. and Duraković, A. (1969) Intestinal absorption of calcium–47 and strontium–85 in lactating rats, *Calcif. Tiss. Res.*, **4**, 13–19.

Kostial, K. and Momčilović, B. (1972) The effect of lactation on the absorption of ^{203}Pb and ^{47}Ca in rats. *Health Phys.*, **23**, 383–4.

Krasowskii, G. N. (1976) Extrapolation of experimental data from animals to man, *Health Perspect.*, **13**, 51–8.

Maxfield, M. E., Stopps, G. J., Barnes, J. R. *et al.* (1972) Effect of lead on blood regeneration following acute hemorrhage in dogs. *Am. Ind. Hyg. Assoc. J.*, **37**, 326–37.

Muraoka, Y. and Itoh, F. (1980) Sex difference of mercuric chloride-induced renal tubular necrosis in rats. *J. Toxicol. Sci.*, **5**, 203–14.

Oliver, T. (1902) *Dangerous Trades*, John Murray, London, p. 296.

Parizek, J. (1956) Effect of cadmium state on testicular tissue. *Nature*, **177**, 1036–7.

Pence, D. H., Miya, T. S. and Schnell, R. C. (1977) Cadmium alternation of hexobarbital action: sex-related differences in the rat. *Toxicol. Appl. Pharmacol.*, **39**, 89–96.

Piasek, M. and Kostial, K. (1987) Effect of exposure to lead on reproduction in male rats. *Bull. Environ. Contam. Toxicol.*, **39**. 448–52.

Piasek, M., Maljković, T., Blanuša, M. *et al.* (1989) Sex-related differences in hemopoietic response to lead in rats. in *Proceedings of the 7th International Conference 'Heavy Metals in the Environment'* (ed. J. P. Vernet), Vol. 1, Geneva, pp. 382–5.

Rabar, I. (1976) Influence of some factors on manganese metabolism in rats, MSc thesis, University of Zagreb, Yugoslavia (in Croatian).

Sarić, M., Prpić-Majić, D., Kostial, K. and Piasek, M. (1987) Exposure to lead and reproduction, in *Summary Proceedings of a Workshop 'Selected Aspects of Exposure to Heavy Metals in the Environment: Monitors, Indictors and High Risk Groups'*, National Academy Press, Washington DC, pp. 119–26.

Schnell, R. C., Pence, D. H., Prosser, T. D. and Miya, T. S. (1976) Sex-related differences in cadmium-induced alternation of drug action in the rat. *Toxicology*, **6** 277–9.

Stowe, H. D. and Goyer, R. A. (1971) The reproductive ability and progeny of F1 lead toxic rats, *Fertil. Steril.*, **22**, 755–60.

Stuik, E. J. and Zielhuis, R. L. (1975) Increased susceptibility of females to inorganic lead, in *Proceedings of the International Symposium on Recent Advances in the Assessment of the Health Effects of Environmental Pollution*, Vol. 2, Commission of the European Communities, Luxembourg, pp. 537–43.

Webb, M. (1972) Binding of cadmium ions by rat liver and kidney. *Biochem. Pharmacol.*, **21** 2751–65.

WHO (1984) *Guidelines for Drinking-Water Quality, Vol. 1 Recommendations*, World Health Organization, Geneva.

WHO (1987) *Air Quality Guidelines for Europe*, WHO regional publication, European series, No. 23, WHO Regional Office for Europe, Copenhagen.

Wibowo, A. A. E. and Zielhuis, R. L. (1980) Different effects on hemesynthesis in male and female rabbits treated with lead acetate. *Arch. Toxicol.*, **45**, 67–73.

PART FOUR

Pharmacogenetic Evidence

8 Lessons from pharmacogenetics

W. Kalow

Department of Pharmacology, University of Toronto, Toronto, Ontario, Canada

8.1 SUMMARY

Pharmacogenetics deals with inborn variation in the response to drugs, ecogenetics with that to chemicals in general; all that has ever been learned through pharmacogenetics applies to ecogenetics.

Historically, pharmacogenetics arose from the discovery of major effects of single genes upon certain drug responses. Later work showed that it was multifactorial inheritance which regularly determined or affected the human capacity for drug elimination; if drug elimination is multifactorial, specific metabolic reactions may still be monogenic.

Past impediments to the discovery of polymorphic (= frequently occurring monogenic) variants have included an overemphasis on the kinetics rather than the chemistry of drug elimination, and the testing of insufficient numbers of subjects.

In most areas of pharmacogenetics, functional studies are still a necessity and cannot yet be discarded in favour of molecular genetic investigations. In addition, for interindividual differences, pharmacogenetics accounts for many well-documented interethnic differences. Hence, the scope and impact of ethnic peculiarities upon the response to toxicants deserves to be illuminated by further investigations. Comparisons between populations will be facilitated by standardization of the methods which have proved useful in the study of single populations. As well as the epidemiological methods which rely on the investigation of people, in vitro studies with human cells or tissues should facilitate new discoveries and comparisons.

8.2 INTRODUCTION

Ecogenetics (Calabrese, 1984; Goedde and Lentner, 1986) is a branch of science concerned with hereditary factors which determine the pharmacological or toxicological responses to chemicals, be these drugs, food constituents or natural or man-made components of the environment (Omenn and Gelboin, 1984; Woodhead *et al.*, 1988). To the extent that pharmacogenetics (Vesell, 1987; Kalow and Spielberg, 1989; Szorady, 1989) is concerned only with drugs, the difference between this discipline and ecogenetics is merely one of scale; the same biochemical defence systems protect us from toxic injury by most chemicals including drugs. However, any idiosyncrasy of a person or a population towards a pollutant in air or water is a different matter from an idiosyncrasy towards a drug. Exposure to a drug can usually be avoided or wilfully terminated by those adversely affected, while the pollutant may inescapably engulf its victim. On the other hand, studies which involve the deliberate application of chemicals to humans will always be, for ethical reasons, in the realm of pharmacogenetics. Thus, idiosyncrasies to drugs are more easily studied than those to non-drug chemicals. Hence I will have to quote from pharmacogenetics even when the implications spell ecogenetics.

8.3 A POINT OF DEPARTURE: MONOFACTORIAL (MENDELIAN) INHERITANCE

In 1956, I found a genetic variant of plasma cholinesterase (Kalow, 1956). I was a pharmacologist who had landed in the field of genetics. When I had measured the effects of succinylcholine in persons with different levels and forms of cholinesterase activity (Kalow and Gunn, 1957), I searched the literature to find out whether there were other examples of the dependence of drug responses on genetic factors. The most important cases which I encountered were the then recent reports on primaquine-induced haemolysis, now referred to as G–6-PD deficiency (Luzzatto, 1986), and on the discrimination between fast and slow metabolizers of isoniazid, a distinction now referred to as acetylation polymorphism (Weber, 1987). There were a few additional cases but none as consequential as these three. While I tried to describe all the findings in a monograph (Kalow, 1962), Denborough and Lovell (1960), in Australia, reported a case of what we now call malignant hyperthermia (Britt, 1987).

In short, Mendelian traits of pharmacogenetic significance have been known for well over 30 years. While awareness of these cases is widespread in the affected medical specialties, these observations have not widely affected the practice of medicine.

8.4 THE DISCOVERY OF MULTIFACTORIAL INHERITANCE IN PHARMACOGENETICS

Following this initial series of observations in the 1950s there was a pause in discovery, that is no further monogenic defects with a strong influence on human drug response were detected for about a decade. Pharmacogenetics seemed to have reached a stage of maturity during which the previously discovered defects were further explored (Kalow, 1984a).

During this time, I heard a comment on this pause in discovery which I never forgot. It was in a talk by Arno Motulsky in which he made a remark to the effect that most drug responses seem to have such a complex basis that monogenic defects may be only rare contributors to pharmacogenetic variation. The truth and the importance of this thought was not generally recognized at that time. This concept is complementary to the notion that each of many gene products may influence a drug's fate or action, so that the sum of such influences constitute **multifactorial** genetic control (Vogel and Motulsky, 1986) of the response to that drug. Multifactorial genetic control is measured in terms of heritability.

Studies in twins which showed high heritability of the elimination rates of such different drugs as dicoumarol (Vesell and Page, 1968a), phenylbutazone (Vesell and Page, 1968b) and antipyrine (Vesell and Page, 1968c) were the bright events during the 1960s. These observations kept clinical interest in pharmacogenetics alive and stimulated other investigators to do similar tests. The fate of virtually all drugs tested in twin studies showed a high heritability (Vesell, 1985); this implies that the fate of most non-drug chemicals will also be genetically influenced or controlled.

Perhaps it was the success of these observations into the fate of drugs that directed many subsequent studies in pharmacogenetics towards the assessment of heritability of drug metabolism, rather than heritability of drug response. In any case, the metabolic fate of drugs and other chemicals always differs more or less between people.

In short, the presence of multifactorial genetic determinants of the human capacity to metabolize drugs and other foreign chemicals is the rule rather than the exception; the presence of such determinants can hide the contribution of a major gene, and it does not exclude the operation of environmental modifiers.

8.5 KINETIC, METABOLIC OR MOLECULAR STUDIES?

Stimulated by Vesell's twin studies, my colleagues and I searched for and found a high heritability of the elimination rate constant of amobarbital (Endrenyi *et al.*, 1976). I soon wondered what we had achieved with this finding. What does 'heritability' of a rate constant mean biologically? The effort to give biochemical meaning to this heritable pharmacokinetic constant of amobarbital turned out to cost several years' work which brought the discovery of glucosidation as a new metabolic pathway (Tang *et al.*, 1978, 1979); amobarbital glucoside in urine usually accounted for 10–30% of the dose, hydroxyamobarbital being the main metabolite. As a result of a genetic defect, amobarbital glucoside was missing in some subjects (Kalow *et al.*, 1977). It was instructive to note that the absence of glucosidation would have escaped detection if we had relied only on measuring the elimination rate of the parent drug (Kalow *et al.*, 1979). This observation served to impress us with the idea that a genetic study of a drug's fate should include measurements of metabolites.

By measuring the appearance of a metabolite, one can hope to assay a single enzyme activity, but this hope is not always fulfilled. For instance, ethanol is oxidized to acetaldehyde by both alcohol dehydrogenase and by cytochrome P450IIE1 (Liever, 1988). We found at least two cytochromes P–450 capable of hydroxylating amobarbital (Reilly *et al.*, 1983), providing a redundancy which may explain why we have never seen an absence of amobarbital hydroxylation. A very recent example of such redundancy of metabolic capacity pertained to nifedipine and the 6-beta-hydroxylation of various steroids by cytochromes P–450hPCN1 and hPCN3 (Aoyama *et al.*, 1989). Nevertheless, monitoring a given metabolite offers a chance to assess a single gene product.

It was an emphasis on metabolites which led to the definition of the debrisoquine-sparteine polymorphism (Idle and Smith, 1984; Eichelbaum, 1988) and of the independent mephenytoin polymorphism (Wilkinson et al., 1989). In both cases, the discovery rested on the measurement of a drug metabolite in urine. For debrisoquine, the use of a 'metabolic ratio' (the concentration of drug over that of the metabolite in 8-hour urine) was possible and proved to be a mathematically desirable method for assessing the metabolizing capacity of individuals.

During the past few years, the discovery of genetic variation of a drug-metabolizing enzyme was regularly followed by attempts to define the variant at the level of genes and proteins. At the present time, particularly active research takes place in the area of molecular genetics of the cytochrome P–450 system (Guengerich, 1987; Gonzalez, 1989 Nebert *et al.*, 1989), man's most important defence system against chemical injury (Table 8.1). Such research has revealed, for instance, that the deficiency of oxidation of debrisoquine and sparteine represents one of three different intron mutations (Gonzalez *et al.*, 1988) leading to splicing defects and loss or absence of the enzyme (Zanger *et al.*, 1988). Thus, three defects at the molecular level translate into one defect at the clinical level. However, only about 70% of Caucasian subjects with a deficiency of debrisoquine/sparteine oxidation can be recognized currently with RFLP-tests for lymphocyte RNA. Furthermore, it needs to be shown to what extent the test works in other populations, since there has been an indication of ambiguous results among Chinese subjects studied in Stockholm (Yue *et al.*, 1989).

Table 8.1 Human cytochromes P450

Chromosome location	*P450*	*Trivial name*	*Prototype substrates*
15	IA1	P_1	Benzo(a)pyrine
15	IA2	P_3	Phenacetin, caffeine, aflatoxin
	IA2	form 4	
19	IIA3	P450(1)	
	IIA3	IIA3	?
19	IIB6	LM2	
	IIB7	LM2	?
	IIB8	LM2	
10	IIC8	form 1	
	IIC8	mp–12, mp–20	Tolbutamide
	IIC9	IICI	
	IIC10	mp–8	Mephenytoin, mephobarbital
22	IID6	dbl	Debrisoquine, sparteine,
	IID7		codeine
	IID8		
10	IIE1	j	Ethanol, halothane, *N*-nitrosodimethylamine
7	IIIA3	HLp	Various steroids, nifedipine,
	IIIA4	nf–25	cyclosporine
	IIIA5	pcn3	
1	IVB1	KLung P450	?
15	XIA1	scc	Cholesterol
10	XVIIA1	17	Various steroids
15	XIXA1	arom	
6	XXIA2	(pseudogene c21A) c21B	

Compiled from Guengerich, 1987; Gonzalez, 1989; Nebert *et al.*, 1989.

In short, studies designed to assess or measure pharmacokinetic variation promise to be most informative when metabolites are measured besides parent drug. While recombinant methodology to study the molecular biology of drug metabolizing enzymes is advancing rapidly, functional studies in man are still a necessity and probably will remain so for a long time to come. Studies of 'population pharmacokinetics' based on clinical data tended in the past to omit metabolite measurements (for example Antal *et al.*, 1989) but perhaps could be redesigned to facilitate the detection of genetic polymorphisms.

8.6 THE SCALE OF INVESTIGATIONS

The term genetic polymorphism refers to the occurrence of genetic variants in a population, whereby the occurrence must be more frequent than might be accounted for by new mutation (Vogel and Motulsky, 1986). A frequently used arbitrary guide is the occurrence of a variant with an allele frequency of 1–2%. At that level of frequency, one-quarter to one-third of all enzymes are found to be genetically variable by functional or physicochemical criteria. Furthermore, enzymes acting on environmental substrates may be more variable than those metabolizing endogenous ones. The clinical significance of many enzyme variants may be limited, particularly in heterozygotes. Nevertheless, genetic variation among the human drug metabolizing enzymes should be fairly widespread, and we can expect to discover many more variants than are known today.

If we wanted to detect enzyme variants that occur in 1% of the population, i.e., in about 2.5 million people in the USA and Canada, we should involve 300–500 subjects in each test that we perform for drug metabolizing capacity. This is a far greater number than used in customary testing of drug metabolism which is mostly done for the purpose of defining the metabolic fate of a drug. This purpose and the purpose of defining variability of drug metabolism in populations have different requirements. Most current investigations involve not only too few subjects for toxicologists to have a good chance of seeing new monogenic variables, but there is also a tendency to neglect outliers by describing population data merely in terms of averages and standard deviations. Investigations of old (i.e. established) monogenic variants are a different matter. According to Ellard (1976), there had been in 1976 as many as 7274 subjects investigated in studies of the acetylation polymorphism; a similar number of subjects has been tested already for the debrisoquine/sparteine oxidation deficiency (Steiner, 1987). Investigations of cholinesterase polymorphism which can be performed with samples of serum or plasma (Whittaker, 1986) probably number one hundred thousand.

If investigations require the testing of metabolite formation in numerous subjects, test methods have to be simple and cheap. Tests requiring blood samples will rarely be suitable. Saliva has been successfully used to obtain pharmacokinetic parameters, as for instance with antipyrine, but metabolite studies of urine appear to be most promising.

Urine collections following the intake of a test drug can be conducted in various ways (Tang *et al.*, 1982, 1983). Sequential sampling of urine may allow considerable insight into the various processes governing the fate of a drug in the body. A substantial reduction of sampling may sometimes lead to only a small loss of information. The extreme simplification is to collect one spot sample of urine per person at a set time after drug intake and to determine the concentration of drug and drug metabolites in it (Grant *et al.*, 1984). Results expressed in terms of concentration ratios may be stable parameters which can be useful for mass screening. This has been abundantly shown in the numerous studies with debrisoquine and sparteine (Jackson *et al.*, 1986; Henthorn *et al.*, 1989). In our earlier studies with amobarbital (Kalow *et al.*, 1978), we were surprised that in a not very rigidly standardized test, not only the metabolite ratio but also the concentration of several metabolites taken separately appeared to give meaningful data on the distribution of drug metabolizing capacities in different populations.

We now realize that, when the purpose of a study is to assess variability patterns in populations, maximal simplification of experimental procedures is appropriate. The information which is thereby provided pertains to human biology and is of potential use to industry and to regulatory agencies. When the metabolic classification of an individual could have medical consequences for that individual, simplified tests may not make sense. The optimization of population surveys deserves new planning and new thoughts, and needs to be guided by clear concepts of purpose.

In short, current knowledge of the distribution of drug-metabolizing capacities in populations is very limited for historical reasons. For remedy, studies are called for which take epidemiological, pharmacokinetic, genetic and medical considerations into account.

8.7 FROM GENETICS TO ANTHROPOLOGY

The difference between our volunteers of Caucasian and Oriental background with respect to amobarbital elimination was a completely unexpected observation, brought about by the ethnic diversity of the student population of the University of Toronto (Kalow *et. al.*, 1979). This observation alerted us to the possibility that interethnic differences of drug metabolizing capacity may be widespread, in contrast to what is generally assumed (Kalow, 1982). We therefore conducted further studies using different drugs and we found interethnic differences for metabolites of caffeine (Kalow, 1985) and of debrisoquine (Kalow, 1984d), but not of antipyrine (Inaba *et al.*, 1981). I tried to give a name to this field of interdisciplinary investigation by calling it 'pharmacoanthropology' (Kalow, 1984b and Kalow, 1984c). I will cite in the following paragraphs some examples which indicate the great variety of pharmacogenetic and ecogenetic differences (Kalow *et al.*, 1986) between representatives of the three major human races ('Caucasoids', 'Mongoloids' and 'Negroids').

8.7.1 Pharmacoanthropology and drug metabolism

1. Debrisoquine/sparteine monooxygenase (P450IID6) deficiency.
This deficiency is clinically important for several series of drugs which include antidepressants, neuroleptics and antiarrhythmics and beta-blockers. The deficiency has a frequency of about 7% (5–10%) in European or European-derived populations but is more rare in all other tested populations (Eichelbaum, 1988), with the exception of San-Bushmen (Sommers *et al.*, 1988). The deficiency has somewhat different characteristics in Europeans and in Ghanaians (Eichelbaum and Woolhouse, 1985). The deficiency appears to be present in less than 1% of Chinese (Lou *et al.*, 1987) and Japanese (Horai *et al.*, 1989) subjects but there is a tendency for a low average activity of the 'normal' (= extensive metabolizer) enzyme.

Tests with microsomal preparations of human liver suggest that there are also variants of the non-deficient form of the enzyme (Tyndale *et al.*, 1989). This could be important since an excess occurence of bronchial carcinoma was observed mainly in persons with very high debrisoquine hydroxylase activity (Ayesh *et al.*, 1984; Caporaso *et al.*, 1989). Furthermore, there might be variants of the active enzyme predominating in some populations so as to account for some of the interethnic differences in the enzyme activity.

2. Mephenytoin hydroxylase deficiency
Mephentoin is a rarely used antiepilepsy drug but it led to the discovery of the deficiency of cytochrome P450IIC8 (Wilkinson *et al.*, 1989). In subjects with the deficiency, the metabolism of diazepam is impaired (Bertilsson *et al.*, 1989). The deficiency is present in about 20% of Oriental but in only 3% of Caucasian populations (Jurima *et al.*, 1985; Nakamura *et al.*, 1985); this implies by the Hardy-Weinberg law that only about 30% of Orientals, against almost 70% of Caucasians, are homozygotes with full enzyme capacity. The difference in customary dose of diazepam (Rosenblat and Tang, 1987) may be explained by this polymorphism; other consequences will surely be discovered.

3. A variant of alcohol dehydrogenase which is uncommon in Europe is the common form of that enzyme in East Asia (Ikuta *et al.*, 1988). The difference is unimportant for ethanol but in vitro work (Kassam *et al.*, 1989) suggests that it could be important for the toxicity of other industrially used alcohols (Table 8.2).

Table 8.2 Kinetic parameters for the oxidation of various alcohols (mean ± SD) by the Caucasian (beta–1) and Oriental (beta–2) types of hepatic alcohol dehydrogenase

Substrate	*Variable*	*Alcohol dehydrogenase type*		
		beta–1ADH (n =4)	*beta–2ADH (n = 4)*	*P*
Ethanol	K_m	255 ± 53	134 ± 25	0.0001
	V_{max}	6.9 ± 4.9	15.2 ±7.9	0.0131
Nicotinyl alcohol	K_m	653 ±127	682 ± 223	NS
	V_{max}	23.8 ±3.3	13.2 ±6.8	0.0001
Benzyl alcohol	K_m	50 ± 27	85 ± 30	0.0002
	V_{max}	17.2 ±3.8	9.0 ±3.9	0.0001

Data compiled from Kassam *et al.*, 1989.
Three determinations were made for each of the eight livers examined.
K_m expressed in μM.
V_{max} expressed in μmoles NADH $min^{-1}mg\ prot^{-1}$.

4. The absence of a form of aldehyde dehydrogenase in 40–50% of Oriental subjects causes a flushing response on ethanol intake, but it also protects against alcoholism (Goedde *et al.*, 1985).

5. The frequency of subjects who are homozygous for the atypical plasma cholinesterase is in the order of 1:4000 in Caucasoids, 1:40 000 in Negroids and 1:400 000 in Mongoloids (Antal *et al.*, 1989). Among the many substrates of the enzyme are aspirin (Rainsford *et al.*, 1980), cocaine (Steward *et al.*, 1979) and heroin (Lockridge *et al.*, 1980).

6. Paraoxonase is an arylesterase of human serum with the ability to hydrolyse the organophosphorous compound paraoxon and other insecticidal or neurotoxic chemicals (LaDu, 1988). Human beings can be classified as having either high or low paraoxonase activity in their blood. The toxicological significance of this difference is as yet uncertain. Nevertheless, there could be ethnic differences in the susceptibility to toxic damage by certain organophosphates since the proportion of subjects with high or low activity varies considerably between ethnic groups.

7. *N*-acetyltransferase polymorphism is affected by geography (Weber, 1987) in that there is a trend for a decreased proportion of slow acetylators at increased latitudes. The acetylation capacity affects the metabolism not only of drugs like isoniazid but also that of arylamine carcinogens like benzidine (Hein, 1988). The activities of *N*-acetyltransferase ('NAT') and *N,O*-arylhydroxamic acid acyltransferase ('AHAT') appear to be due to the same enzyme, but it has become clear that there is another closely related acetyltransferase capable of acting upon aromatic amines (Grant *et al.*, 1989); epidemiological studies based on that knowledge are not yet available.

8.7.2 Pharmacoanthropology and drug action

1. The deficiencies of glucose–6-phosphate dehydrogenase (Luzzatto, 1986) were discovered as a consequence of investigations into the primaquine-induced haemolysis of black but not white American soldiers. The uneven geographical distribution of the deficiencies is explained by their conveying resistance against falciparum malaria. There are hundreds of enzyme variants of which many are rare; nevertheless, most have been classified according to their medical significance. Approximately 500 million people are affected. There are thousands of relevant publications.

2. The incidence of hypertensive disease, its basic pathogenesis and therefore the optimal drug therapy of hypertension differ between blacks and whites in America and elsewhere (Freis, 1986; Motulsky *et al.*, 1987). One may theorize that populations, genetically adapted to preservation of body salt in hot climates tend to suffer salt-induced hypertension . . . when participating in the liberal salt consumption of Western civilization.

3. A deficiency of alpha$_1$-antitrypsin in homozygotes of a variant allele referred to as '*PI Z*' is virtually confined to populations in and from Northern Europe (Cox, 1986). The homozygotes are rare (1 in 2000 to 8000), but they tend to have a shortened lifespan due to either liver disease or to emphysema produced by smoking or air pollution. The toxicological significance of the various heterozygosities in 3–4% of Caucasians is as yet uncertain.

4. In order to explain the low-dose requirements for beta-blockers in China, Zhou *et al.* (1989) compared doses, blood levels and circulatory effects of propranolol in Americans of Chinese and European extraction. The Chinese metabolized the drug faster but, for given bloodlevels, they responded much more strongly with cardiac slowing and lowering of blood pressure than did the other volunteers (Kalow, 1989). A subsequent study suggested that the response intensity is a function of receptor density.

In summary, there is overwhelming evidence that pharmacogenetic differences may occur not only between individuals but also between ethnically defined populations, irrespective of the fact that non-genetic differences can occur in addition. The same must be true for ecogenetics; indeed, most of the examples quoted on the previous pages (see examples 1, 3, 6 and 7 in section 8.7.1 and 1, 2, and 3 in section 8.7.2) are known or suspected to be relevant for non-drug chemicals.

8.8 HUMAN TISSUES AS OBJECTS OF STUDY

Any search for differences between people cannot be performed in animals. Hence, human cells or tissues are particularly important tools to students of human variation.

Classic pharmacogenetic investigations using autopsy liver revealed that about 40% of Orientals were without the one isozyme of aldehyde dehydrogenase which has high affinity for acetaldehyde, the toxic oxidation product of ethanol. These in vitro data are clear and virtually without ambiguity (Harada *et al.*, 1978), a fact that greatly facilitates research into the clinical meaning of these variations.

The possibilities of in vitro testing of human tissues, cells, cell constituents or tissue fluids are numerous and can only be touched upon (Kalow, 1986). The accessibility of blood makes it a particularly valuable source of materials, containing nucleated white cells and platelets and offering erythrocytes and plasma as enzyme sources (Weinshilboum, 1988) and plasma proteins as binding agencies and carriers. Also hair roots (Goedde *et al.*, 1980) and skin fibroblasts have served as objects of pharmacogenetic investigations. There is no theoretical limit in studying tissues obtained at autopsy, by biopsy or as surplus materials from organ transplantations. Investigations might consist of analysis of DNA or RNA or of measurements of enzyme activity or of binding to enzymes, receptors or carrier proteins.

Two questions should be raised, particularly in respect to interethnic comparisons. First, what should be done to standardize and make comparable the tests performed in different laboratories and different countries? Second, should deliberate efforts be made to stimulate the parallel conduction of particular tests in different countries? The first question might be answered by referring to the experiences with the standardization of tests adopted for clinical chemistry in many hospital laboratories. The answer to the second question depends mainly on how we evaluate the different responses of populations in the past to drugs and chemicals, and how we extrapolate to the future. A factor which favours positive answers to both questions is the fact that recent migrations have created many multi-ethnic communities, so that interethnic differences are no longer merely a matter of geography.

In short, it is rational to use human tissues, cells and cell constituents to define in vitro those biochemical factors which affect human metabolizing capacity or response to chemicals, and which differ between individuals or between populations. Encouragement and standardization of such tests are called for.

8.9 CONCLUSIONS

Further research in ecogenetics should not only be aimed at clarifying interindividual differences, but also interethnic differences. The more that is known about the presence or absence of genetic differences in response to drugs and toxicants, the better will be the basis for good policy decisions.

There appears to be a need for new investigations in two directions. One is to study drug metabolizing capacity in populations, thereby providing a database for biochemical epidemiology and correlation studies. For the purpose, assessment of particular metabolites of probe drugs in urine is a proven method; as research progresses, recombinant methodology will offer new and additional opportunities.

The second direction is to increase and systematize the comparative study of human tissues, cells or body fluids as a means of defining and exploring differences between individuals and between populations. An international exchange of specimens and/or standardization of methods is a desirable goal.

REFERENCES

Antal, E. J. Grasela Jr, T. H. and Smith, R. B. (1989). An evaluation of population pharmacokinetics in therapeutic trials. Part III. Prospective data collection versus retrospective data assembly. *Clin. Pharmacol. Ther.*, **46**, 552–9.

Aoyama, T., Yamano, S., Waxman, D. J. *et al.* (1989), Cytochrome P–450 hPCN3, a novel cytochrome P–450 IIIA gene product that is differentially expressed in adult human liver. *J. Biol. Chem.*, **264**, 10388–95.

Ayesh, R., Idle, J. R., Ritchie, J. C. *et al.* (1984). Metabolic oxidation phenotypes as markers for susceptibility to lung cancer. *Nature*, **312**, 169–70.

Bertilsson, L., Henthorn, T. K., Sanz, E. *et al.* (1989) Importance of genetic factors in the regulation of diazepam metabolism: relationship to *S*-mephenytoin, but not debrisoquin, hydroxylation phenotype. *Clin. Pharmacol. Ther.* (1987), **45**, 348–55.

Britt, B. A. (ed.) (1987) *Malignant Hyperthermia*, Martinus Nijhoff, Boston.

Calabrese, E. J. (1984) *Ecogenetics: Genetic Variation in Susceptibility to Environmental Agents*, John Wiley, New York.

Caporaso, N., Hayes, R. B., Dosemici, M. *et al.* (1989) Lung cancer risk, occupational exposure, and the debrisoquine metabolic phenotype. *Cancer Res.* **49**, 3675–9.

Cox, D. W. (1986) Clinical and molecular studies of alpha$_1$-antitrypsin deficiency, in *Ethnic Differences in Reactions to Drugs and Xenobiotics* (eds W. Kalow, H. W. Goedde and D. P. Agarwal), Alan R. Liss, New York, pp. 373–840.

Denborough, M. A. and Lovell, R. R. H. (1960) Anaesthetic deaths in a family. *Lancet* **11**, 45.

Eichelbaum, M. (1988) Genetic polymorphism of sparteine/debrisoquine oxidation. *ISI Atlas Sci.*, **2**, 243–51.

Eichelbaum, M. and Woolhouse, N. M. (1985) Inter-ethnic difference in sparteine oxidation among Ghanaians and Germans. *Eur. J. Clin. Pharmacol.*, **28**, 79–83.

Ellard, G. A. (1976) Variations between individuals and populations in the acetylation of isoniazid and its significance for the treatment of pulmonary tuberculosis. *Clin. Pharmacol. Ther.*, **19**, 610–25.

Endrenyi, L., Inaba, T. and Kalow, W. (1976) Genetic study of amobarbital elimination based on its kinetics in twins. *Clin. Pharmacol. Ther.*, **20**, 701–14.

Freis, E. D. (1986) Antihypertensive agents, in *Ethnic Differences in Reactions to Drugs and Xenobiotics* (eds W. Kalow, H. W. Gredde and D. P. Agarwal), Alan R. Liss, New York, pp. 313–22.

Goedde, H. W., Agarwal, D. P., Eckey, R. and Harada, S. (1985) Population genetic and family studies on aldehyde dehydrogenase deficiency and alcohol sensitivity. *Alcohol* **2**, 383–90.

Goedde, H. W., Agarwal, D. P. and Harada, S. (1980) Genetic studies on alcohol metabolizing enzymes: detection of isozymes in human hair roots. *Enzyme*, **25**, 281–6.

Goedde, H. W. and Lentner, C. (1986) *Pharmacogenetics and Ecogenetics. Geigy Scientific Tables*, 8th edn, Vol. 4, Basle, pp. 289–300.

Gonzalez, F. J. (1989) The molecular biology of cytochrome P450s. *Pharmacol. Rev.* **40**, 243–88.

Gonzales, F. J., Skoda, R. C., Kimura, S. *et al.* (1988) Characterization of the common genetic defect in humans deficient in debrisoquine metabolism. *Nature*, **331**, 442–6.

Grant, D. M., Lottspeich, F. and Meyer, U. A. (1989) Evidence of two closely related isozymes of arylamine *N*-acetyltransferase in human liver. *FEBS Lett*, **244**, 203–7.

Grant, D. M., Tang, B. K. and Kalow, W. (1984) A simple test for acetylator phenotype using caffeine. *Br. J. Clin. Pharmac.* **17**, 459–64.

Guengerich, F. P. (1987) Cytochrome P–450 enzymes and drug metabolism. *Prog. Drug Metab.* **10**, 1–54.

Harada, S., Agarwal, D. P. and Goedde, H. W. (1978) Isozyme variations in acetaldehyde dehydrogenase (EC 1.2.1.3) in human tissues. *Hum. Genet.*, **44**, 181–5.

Hein, D. W. (1988) Acetylator genotype and arylamine-induced carcinogenesis. *Biochim. Biophys. Acta.* **948**, 37–66.

Henthorn, T. K., Benitez, J. Avram, M. J. *et al.* (1989) Assessment of the debrisoquin and dextromethorphan phenotyping tests by gaussian mixture distributions analysis. *Clin. Pharmacol. Ther.*, **45**, 328–33.

Horai, Y., Nakano, M., Ishizaki, T. *et al.* (1989) Metoprolol and mephenytoin oxidation polymorphisms in far eastern oriental subjects: Japanese versus mainland Chinese. *Clin. Pharmacol. Ther.* **46**, 198–207.

Idle, J. R. and Smith, R. L. (1984) The debrisoquine hydroxylation gene: a gene of multiple consequence, in *Proc. 2nd World Conf. Clin. Pharmacol. Ther.* (eds L. Lemberger and M. M. Reidenberg), Am. Soc. Pharmacol. Exp. Ther. Baltimore, pp. 148–64.

Ikuta, T., Shibuya, A. and Yoshida, A. (1988) Direct determination of usual (Caucasian-Type) and atypical (Oriental-Type) alleles of the class I human alcohol dehydrogenase–2 locus. *Biochem. Genet.*, **26**, 519–25.

Inaba, T., Uchino, H., Kadar, D. and Kalow, W. (1981) Antipyrine metabolites in two populations. *Res. Commun. Chem. Pathol. Pharmacol.* **32**, 235–44.

Jackson, P. R., Tucker, G. T., Lennard, M. S. and Woods, H. F. (1986) Polymorphic drug oxidation: pharmacokinetic basis and comparison of experimental indices. *Br. J. Clin. Pharmacol.*, **22**, 541–50.

Jurima, M., Inaba, T., Kadar, D. and Kalow, W. (1985) Genetic polymorphism of mephenytoin p(4′)-hydroxylation: difference between Orientals and Caucasians. *Br. J. Clin. Pharmacol.*, **19**, 483–7.

Kalow, W. (1956) Familial incidence of low pseudocholinesterase level. *Lancet*, **ii**, 576.

Kalow, W. (1962) *Pharmacogenetics. Heredity and the Response to Drugs*. Saunders, Philadelphia.

Kalow, W. (1982) Ethnic differences in drug metabolism. *Clin. Pharmacokinet.*, **7**, 373–400.

Kalow, W. (1984a) A pharmacologist looks at ecogenetics, in*Genetic Variability in Response to Chemical Exposure*(eds. G. S. Omenn and H. V. Gelboin), Banbury Report 16, Cold Spring Harbor Laboratory, pp. 15–32.

Kalow, W. (1984b) Pharmacoanthropology: drug metabolism. *Fed. Proc.* **43**, 2326–31.

Kalow, W. (1984c) Pharmacoanthropology: outline, problems, and the nature of case histories. *Fed. Proc.* **43**, 2314–18.

Kalow, W. (1984d) Pharmacogenetics and anthropology, in *Proc. 2nd World Conf. Clin. Pharmacol. Ther.* (eds L. Lemberger and M. M. Reidenberg), Am. Soc. Pharmacol. Exp. Ther., Baltimore, pp. 264–86.

Kalow, W. (1985) Variability of caffeine metabolism in humans. *Arzneim.-Forsch./Drug Res.*, **35**, 319–24.

Kalow, W. (1986) Prospective pharmacogenetics, in *Proc. Clin. Pharmacol. Symp. II* (eds G. M. Fulgraff, H. Lenau, H. Maier-Lenz and H. Rode), Titisee, pp. 100–16.

Kalow, W. (1989) Race and therapeutic drug response. *N. Engl. J. Med.*, **320**, 588–90.

Kalow, W., Endrenyi, L., Inaba, T. *et al.* (1970) Pharmacogenetic investigation of amobarbital disposition, in *Advances in Pharmacological Therapy Vol. 6, Clinical Pharmacology: Proceedings of the International Congress of Pharmacology, Paris, 1978* (ed. P. Duchene-Marullaz), Pergamon Press, Oxford, pp. 31–40.

Kalow, W., Goedde, H. W. and Agarwal, D. P. (eds) (1986) *Ethnic Differences in Reactions to Drugs and Xenobiotics*, Alan R. Liss, New York.

Kalow, W. and Gunn, D. R. (1957) The relation between dose of succinylcholine and duration of apnea in man. *J. Pharmacol. Exp. Ther.*, **120**, 203–14.

Kalow, W., Kadar, D., Inaba, T. and Tang, B. K. (1977) A case of deficiency of *N*-hydroxylation of amobarbital. *Clin. Pharmacol. Ther.*, **21**, 530–5.

Kalow, W. and Spielberg, S. (1989) Human pharmacogenetics, in *Principles of Medical Pharmacology* (eds H. Kalant and W. Koschlau), 5th edn. B. C. Decker, Burlington, pp. 119–27.

Kassam, J. P., Tang, B. K., Kadar, D. and Kalow, W. (1989) In vitro studies of human liver alcohol dehydrogenase variants using a variety of substrates. *Drug Metab. Dispos.*, **17**, 567–72.

LaDu, B. N. (1988) Invited editorial: The human serum paraoxonase/arylesterase polymorphism. *Am. J. Hum. Genet.*, **43**, 227–9.

Lieber, C. S. (1988) Biochemical and molecular basis of alcohol-induced injury to liver and other tissues. *N. Engl. J. Med.*, **319**, 1639–50.

Lockridge, O., Mottershaw-Jackson, N., Eckerson, H. W. and La Du, B. N. (1980) Hydrolysis of diacetylmorphine (Heroin) by human serum cholinesterase. *J. Pharmacol. Exp. Ther.*, **215**, 1–8.

Lou, Y., Ying, L., Bertilsson, L. and Sjoqvist, F. (1987) Low frequency of slow debrisoquine hydroxylation in a Native Chinese population. *Lancet*, **ii**, 852–3.

Luzzatto, L. (1986) Glucose–6-phosphate dehydrogenase and other genetic factors interacting with drugs, in *Ethnic differences in Reactions to Drugs and Xenobiotics* (eds W. Kalow, H. W. Goedde and D. Agarwal), Alan R. Liss, New York, pp. 385–99.

Motulsky, A. G., Burke, W., Billings P. R. and Ward, R. H. (1987) Hypertension and the genetics of red cell membrane abnormalities, in *Molecular Approaches to Human Polygenic Disease*, Ciba Foundation Symp. 130, John Wiley, Chichester, pp. 150–66.

Nakamura, K., Goto, F., Ray, W. A. *et al.* (1985) Interethnic differences in genetic polymorphism of debrisoquin and mephenytoin hydroxylation between Japanese and Caucasian populations. *Clin. Pharmacol. Ther.*, **38**, 402–8.

Nebert, D. W., Nelson, D. R., Adesnik, M. *et al.* (1989) The P450 superfamily: updated listing of all genes and recommended nomenclature for the chromosomal loci. *DNA*, **8**, 1–13.

Omenn, G. S. and Gelboin, H. V. (eds) (1984) *Genetic Variability in Responses to Chemical Exposure*, Banbury Report 16, Cold Spring Harbor Laboratory.

Rainsford, K. D. Ford, N. L. V., Brooks, P. M. and Watson, H. M. (1980) Plasma aspirin esterases in normal individuals, patients with alcoholic liver disease and rheumatoid arthritis: characterization and the importance of the enzymic components. *Eur. J. Clin. Invest.*, **10**, 413–20.

Reilly, P. A., Tang, B. K., Stewart, D. J. and Kalow, W. (1983) The occurrence of two hepatic microsomal sites for amobarbital hydroxylation. *Can. J. Physiol. Pharmacol.*, **61**, 67–71.

Rosenblat, R. and Tang, W. A. (1987) Do Oriental psychiatric patients receive different dosages of psychotropic medication when compared with occidentals? *Can. J. Psychiatry*, **32**, 270–4.

Sommers, D. K., Moncrieff, J. and Avenant, J. (1988) Polymorphism of the 4-hydroxylation of debrisoquine in the San Bushmen of Southern Africa. *Hum. Toxicol.* **7**, 273–6.

Steiner, E. (1987) Polymorphic debrisoquine hydroxylation, PhD thesis, Dept Clin Pharmacol, the Karolinksa Inst., Huddinge Hospital, Sweden.

Steward, D. J., Inaba, T., Lucassen, M. and Kalow, W. (1979) Cocaine metabolism: cocaine and norcocaine hydrolysis by liver and serum esterases. *Clin. Pharmacol. Ther.* **25**, 464–8.

Szorady, I. (1989) *Klinische Pharmakogenetik* (*Clinical Pharmacogenetics*), Ernst Reinhardt, München (Munich).

Tang, B. K., Kalow, W., Endrenyi, L. and Chan, F. Y. (1982) An assessment of short-cut procedures for studying drug metabolism in vivo using amobarbital as a model drug. *Eur. J. Clin. Pharmacol.*, **22**, 229–33.

Tang, B. K., Kalow, W. and Grey, A. A. (1978) Amobarbital metabolism in man: *N*-glucoside formation. *Res. Commun. Chem. Pathol Pharmacol.*, **21**, 45–53.

Tang, B. K., Kalow, W. and Grey, A. A. (1979) Metabolic fate of phenobarbital in man: *N*-glucoside formation. *Drug Metab. Dispos.*, **7**, 315–18.

Tang, B. K., Kalow, W., Inaba, T. and Kadar, D. (1983) Variation in amobarbital metabolism: evaluation of a simplified population study. *Clin. Pharmacol. Ther.*, **34**, 202–6.

Tyndale, R. F., Inaba, T. and Kalow, W. (1989) Evidence in humans for variant allozymes of the nondeficient sparteine/debrisoquine monooxygenase (P450IID1) in vitro. *Drug Metab. Dispos.*, **17**, 334–40.

Vesell, E. S. (1985) Clinical pharmacology: a personal perspective. *Clin. Pharmacol. Ther.*, **38**, 603–12.

Vesell, E. (1987) Pharmacogenetic perspectives on susceptibility to toxic industrial chemicals. *Br. J. Ind. Med.*, **44**, 505–9.

Vesell, E. S. and Page, J. G. (1968a) Genetic control of dicumarol levels in man. *J. Clin. Invest.*, **47**, 2657–63.
Vesell, E. S. and Page, J. G. (1968b) Genetic control of drug levels in man: phenylbutazone. *Science*, **159**, 1479–80.
Vesell E. S. and Page, J. G. (1968c) Genetic control of drug levels in man: antipyrine. *Science*, **161**, 72–3.
Vogel, F. and Motulsky, A. G. (1986) *Human Genetics, Problems and Approaches*, 2nd edn, Springer, Berlin.
Weber, W. W. (1987) *The Acetylator Genes and Drug Response*. Oxford University Press, New York.
Weinshilboum, R. (1988) Pharmacogenetics of methylation: relationship to drug metabolism. *Clin. Biochem.*, **21**, 201–10.
Whittaker, M. (1986) *Cholinesterase: Monographs in Human Genetics*, Vol. 11, Karger, Basel.
Wilkinson, G. R., Guengerich, F. P. and Branch, R. A. (1989) Genetic polymorphism of *S*-mephenytoin hydroxylation. *Pharmacol. Ther.* **43**, 53–76.
Woodhead, A. D., Bender, M. A. and Leonard, R. C. (eds) (1988) *Phenotypic Variation in Populations: Relevance to Risk Assessment*, Plenum Press, New York.
Yue, O. Y., Bertilsson, L., Dahl-Puustinen, M. L. *et al.* (1989) Disassociation between debrisoquine hydroxylation phenotype and genotype among Chinese. *Lancet*, **ii**, 870.
Zanger, U. M., Vilbois, F., Hardwick, J. P. and Meyer, U. A. (1988) Absence of hepatic cytochrome P450bufI causes genetically deficient debrisoquine oxidation in man. *Biochemistry*, **27**, 5447–54.
Zhou, H. H., Koshakji, R. P., Silberstein, D. J. *et al.* (1989) Racial differences in drug response. *N. Engl. J. Med.*, **320**, 565–70.

9 Human variability in alcohol and aldehyde metabolism: vulnerability to toxic effects

D. P. Agarwal and H. W. Goedde

Institute of Human Genetics, University of Hamburg, Hamburg, Federal Republic of Germany

9.1 SUMMARY

Numerous naturally occurring alcohol and nonalcohol compounds, as well as many aliphatic and aromatic aldehydes present in alcoholic beverages and foodstuffs, are known to be potent organotoxics and carcinogens. The evidence indicates that pharmacogenetic and ecogenetic factors partly predispose to, or protect against, alcohol and aldehyde toxicity, alcohol abuse and alcoholism.

Alcohol dehydrogenase (ADH) and aldehyde dehydrogenase (ALDH) constitute the major enzymes involved in the degradation of these alcohols and aldehydes respectively. Genetic polymorphism in these enzymes may be responsible for imparting vulnerability to, or protection against, some of the toxic effects of alcohol drinking in different human populations. An abnormal ALDH isoenzyme form, ALDH2, which is widely prevalent in Orientals of Mongoloid origin, causes acute sensitivity to alcohol in Japanese, Chinese and Koreans. Alcohol sensitivity and associated discomfort symptoms lead to a significantly low incidence of alcohol abuse and alcoholism among these populations.

9.2 INTRODUCTION

A number of genetically controlled enzyme polymorphisms may significantly affect individual susceptibility to severe toxic effects of certain chemicals and foreign compounds. In recent years, adequate evidence has accumulated that pharmacogenetic and ecogenetic factors partly predispose to or protect against alcohol and aldehyde toxicity, alcohol abuse and alcoholism. While ADH constitutes the major enzyme responsible for alcohol degradation, ALDH is the principal enzyme that oxidizes various aldehydes. Alcohol abuse, alcoholism and other alcohol-related disorders are some of the most challenging current health problems with far-reaching medical, social and economic consequences. The prevalence and incidence of heavy alcohol consumption are increasing everywhere. Each year, hundreds of billions of dollars are spent on health care costs related to alcoholism in the developed, as well as in the developing, countries.

A series of factors might interact in predisposing or protecting an individual against alcohol toxicity, alcohol abuse and other alcohol-related disorders: including the availability and price of alcohol and the individual's sociocultural, psychological, physiological and biological make-up. Differences in the pattern of alcoholic beverage use in diverse populations clearly indicate that besides environmental contributions, genetic factors may play an important role in vulnerability to alcohol-related disorders. Different drinking norms and beliefs in relation to drinking practices in different populations and ethnic groups may interact with metabolic and biological variables in modulating alcohol abuse and alcoholism.

This chapter presents recent findings concerning genetic variability in alcohol metabolism and its role in detoxification of various alcohol and aldehyde compounds, as well as predisposition to toxic effects of alcohol use and abuse.

9.3 METABOLISM OF ETHANOL, HIGHER ALCOHOLS AND ALDEHYDES

9.3.1 Absorption and elimination of ethanol

After ingestion, ethanol is readily absorbed from the gastrointestinal tract and diffuses rapidly and uniformly throughout the body water. Approximately 90% of the ingested ethanol is eliminated by oxidation to carbon dioxide and water. Small amounts of ethanol are excreted unchanged in expired air, urine and sweat. The total alcohol eliminated by the human body is usually in the range of 100–300 mg/kg per hour, equivalent to about 6–9 g alcohol per hour for a healthy subject with an average body weight. The rate of ethanol elimination from the body is nonlinear, and many pharmacokinetic models have been used to understand the course of ethanol elimination in humans (von Wartburg, 1989).

A striking similarity in ethanol metabolism rate has been observed in identical twins, with much greater variability between fraternal twins (Vesell *et al.*, 1971; Martin *et al.*, 1985). Comparisons of ethanol metabolism rate in different population groups are shown in Table 9.1.

Table 9.1 Rate of alcohol metabolism in different population groups (reproduced from Agarwal and Goedde, 1990, with permission from the publisher)

Population	*Sample size*	*Alcohol metabolism rate (mg/kg per h)*
Caucasians		
Europeans	19	86.4
Europeans	16	108.0
Europeans	23	108.0
Europeans	6	103.0
North Americans	30	93.2
North Americans	37	103.6
Canadians	68	108.0
North Americans	17	112.0
North Americans	17	144.9
Hindu Reddis	35	122.9
Mongoloids		
Chinese	15	136.6
Chinese	39	127.0
Mongoloids (mixed)	24	146.0
Japanese	47	133.6
Japanese	68	119.0 – 138.0
Native Americans		
Canadian Indians	26	101.3
N. American Indians	30	92.4
Ojibwa Indians	12	182.7
N. American Indians	17	122.9
Canadian Eskimos	21	109.8

9.3.2 Biochemical pathways of alcohol degradation

As shown in Fig. 9.1, three principal enzymes – the cytosolic alcohol dehydrogenase (ADH), the microsomal ethanol oxidizing system (MEOS) and catalase – oxidize ethanol to acetaldehyde (Agarwal and Goedde, 1989, 1990). Acetaldehyde is further oxidized to acetate, which is then converted to carbon dioxide via the citric acid cycle.

The MEOS accounts for the major non-ADH pathway of ethanol oxidation in the liver. The enzyme system is NADPH-dependent and is linked to the cytochrome P–450 oxygenases. Chronic ethanol intake leads to enhanced MEOS activity, which in turn may be responsible for the increased rate of alcohol metabolism commonly observed after prolonged ethanol administration. This adaptive increase in ethanol oxidation

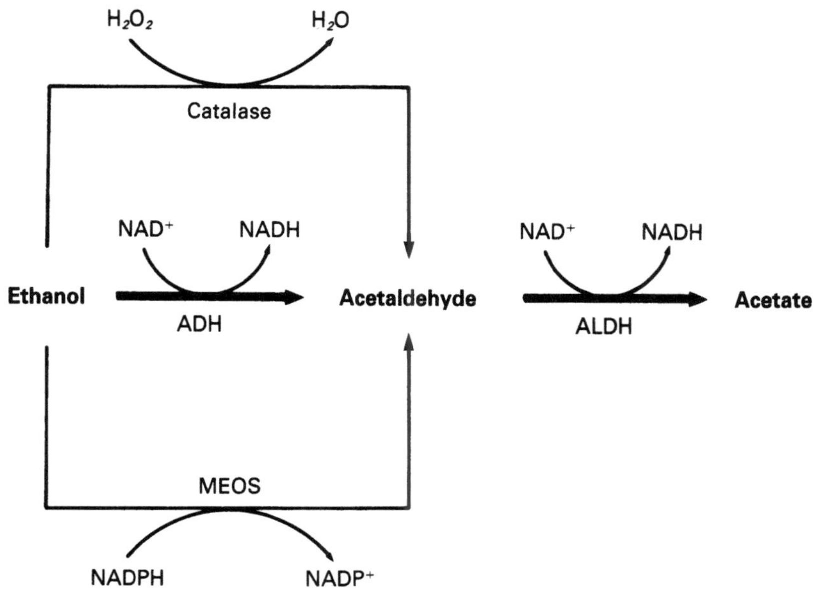

Figure 9.1 Enzymatic pathways for ethanol metabolism (from Agarwal and Goedde, 1990, reproduced with permission of the publisher).

results in enhanced production of acetaldehyde, which exerts a variety of toxic effects at cellular and subcellular level. In addition, induction of MEOS may play an important role in the hepatotoxicity of a number of drugs, carcinogens and xenobiotics by way of an increased metabolism of these compounds. Catalase may also be involved in the oxidation of a variety of alcohols to their corresponding aldehydes in a peroxidase-type reaction. However, the main rate-limiting factor for catalase activity is the rate of hydrogen peroxide formation.

(a) Alcohol dehydrogenase

Liver alcohol dehydrogenase (ADH, alcohol:NAD^+ oxidoreductase, EC 1.1.1.1) constitutes the major oxidative pathway for ethanol metabolism. ADH is universally distributed in living organisms. Human ADH is a dimeric protein consisting of two subunits of molecular weight 40 000 daltons each. At east five different gene loci (ADH_1 to ADH_5) code for human ADH arising from the association of eight different types of subunits: α, β_1, β_2, β_3, γ_1, γ_2, π and χ. The different molecular forms of ADH can be divided into three major classes (I, II and III) according to their isozyme composition and electrophoretic and kinetic properties. Class I isozymes are formed by random dimeric association of any of the three types of polypeptide subunits (α, β and γ) controlled by three separate gene loci, ADH_1, ADH_2 and ADH_3, respectively (Smith *et al.*, 1973). Class I ADH isozymes exhibit a high activity for ethanol oxidation and are considered to play a major role in ethanol metabolism in humans.

An anodic isozyme, π-ADH has catalytic properties significantly different from those of other ADH isozymes (Bosron *et al.*, 1979). The π-ADH encoded by the ADH_4 locus migrates anodically to the class I isozymes on starch gel electrophoresis. This isozyme has a significantly higher K_m value for ethanol (34 mmol/l) and cannot oxidize methanol. Another ADH isozyme controlled by the ADH_5 locus, designated χ-ADH, shows widely different catalytic and structural properties as compared to class I and class II enzymes (Juliá *et al.*, 1987). Only long-chain alcohols, such as pentanol and octanol, and aromatic alcohols, such as cinnamyl alcohol, are oxidized by χ-ADH, and the enzyme is not inhibited by 4-methylpyrazole (Parés and Vallee, 1981).

(b) Alcohol dehydrogenase polymorphisms

While an extensive polymorphism has been observed for the ADH_2 locus, no allelic variation has been reported in human populations for ADH_1 (class I ADH), π-ADH (class II ADH) and χ-ADH (class III ADH). The multiple molecular forms of class I ADH are identified on starch gel electrophoresis or isoelectric focusing in polyacrylamide gels as a complex set of distinct isozymes (Fig. 9.2). Any one particular isozyme may be composed of homodimeric subunits, consisting of two identical polypeptides (e.g. $\alpha\alpha$, $\beta\beta$, $\gamma\gamma$) coded by a specific allele at one of the loci, or heterodimeric subunits, consisting of two nonidentical polypeptides (e.g. $\alpha\beta$, $\beta\gamma$) coded by alleles at separate loci, or heterodimeric subunits, but coded by different alleles at the same locus (e.g. $\beta_1\beta_2$, $\gamma_1\gamma_2$).

Livers containing the normal ADH_2 phenotype ($\beta_1\beta_1$ subunits) exhibit a pH optimum at about 10.8, whereas livers with either an atypical heterozygote ($\beta_1\beta_2$) or an atypical homozygote ($\beta_2\beta_2$) form show a pH optimum at 8.5. This variant enzyme, encoded by the ADH_2 locus and commonly known as the 'atypical' ADH, migrates faster towards the cathode in starch gel on electrophoresis. The homodimeric $\beta_2\beta_2$ variant form has higher K_m values for ethanol and NAD^+ and a higher V_{max} for ethanol than $\beta_1\beta_1$ and $\beta_1\beta_2$ forms (Bosron and Li, 1986).

A new polymorphic form arising from the ADH_2 locus exhibits a dual pH optimum for ethanol oxidation at pH 7.0 and 10.0, and a greater electrophoretic mobility toward the cathode is designated as $ADH_{Indianapolis}$. This variant was not found in Caucasians in North America, Germans nor Japanese (Bosron *et al.*, 1980; Agarwal *et al.*, 1981). As this variant

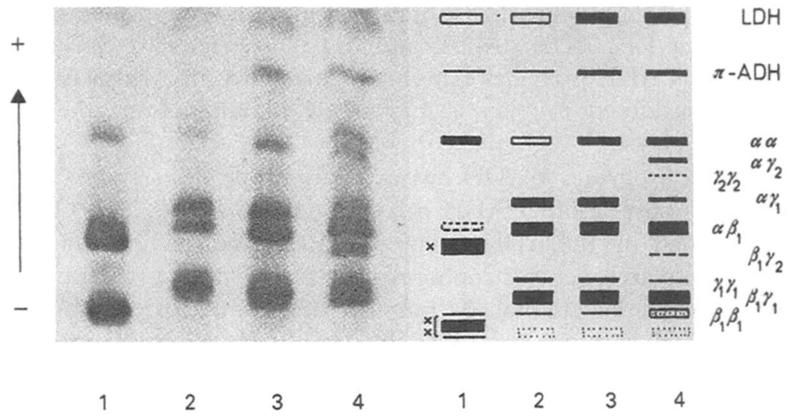

Figure 9.2 Electrophoretic pattern of alcohol dehydrogenase (ADH) isozymes after high-voltage starch gel electrophoresis of liver extracts followed by enzyme activity staining. Lane 1: ADH_2^{2-1}; lanes 2 and 3: ADH_2^{1-1}, ADH_3^{1-1}; lane 4: ADH_2^{1-1}, ADH_3^{2-1}. (From Agarwal and Goedde, 1990, reproduced with permission of the publisher).

is the third type of β subunit that has been characterized it is now designated β_3.

The isozymes determined by the ADH_3 locus are made up of γ polypeptide subunits. ADH_3 is also polymorphic and shows heterodimeric ($\gamma_1\gamma_2$) and homodimeric ($\gamma_2\gamma_2$) allelic forms. The $\gamma_1\gamma_1$ isozyme has a slower electrophoretic mobility than the $\gamma_2\gamma_2$ form. The differences in γ_1 and γ_2 subunits have also been detected at the protein and DNA level.

(c) Frequence of ADH alleles in different populations

About 5–10% of the English, 9–14% of the German and 20% of the Swiss populations have the 'atypical' ADH (ADH_2 locus), whereas this variant form occurs in at least 85% of the Japanese and Chinese (Table 9.2). The $ADH_{Indianapolis}$ variant allele was found to be present in about 25% of the black American populations but not in Caucasians in North America, Germans nor Japanese (Bosron *et al.*, 1980; Agarwal *et al.*, 1981).

Table 9.2 Distribution of atypical ADH phenotype in different populations (reproduced from Agarwal and Goedde, 1990, with permission from the publisher)

Population	*Frequency of atypical ADH (%)*
Swiss	20
English	5–10
Germans	9–14
Japanese	85–98
Chinese	89
Bahia (Brazil)	2.8
Asian Indians	0

The frequencies of ADH_3^1 and ADH_3^2 alleles in Caucasians were reported to be 0.6 and 0.4 respectively (Smith *et al.*, 1972), whereas in the Chinese and Japanese populations the corresponding frequencies of the ADH_3^1 and ADH_3^2 alleles were found to be 0.91 and 0.09 (Teng *et al.*, 1979; Harada *et al.*, 1980a). In mixed populations from Brazil (Bahia) and black American populations the ADH_3^2 allele is relatively less frequent (Azevédo *et al.*, 1975; Bosron *et al.*, 1988). The ADH_3^2 gene frequency was significantly higher (0.55) than the ADH_3^1 gene frequency (0.45) in American Indians from New Mexico (Rex *et al.*, 1985; Bosron *et al.*, 1988). The class I ADH genotypes at the ADH_2 and ADH_3 loci have been determined by the use of allele-specific oligonucleotides and DNA sequences amplified in vitro by the polymerase chain reaction technique (Gennari *et al.*, 1988; Ikuta *et al.*, 1988; Xu *et al.*, 1988). The gene frequencies for different alleles were found to be quite similar to those reported by phenotyping techniques.

(d) Structures of ADH isozymes

All ADH subunits consist of 374 amino acid residues and have about 10% total amino acid exchanges. At three positions (143, 319, 327) all three subunits differ from each other. The β_2 subunit (atypical ADH form) undergoes an Arg–47 to His–47 exchange, which explains a lower pH optimum, a higher turnover number and a lower affinity for coenzyme binding (Jörnvall *et al.*, 1987). The Arg/Gln exchange at position 271 in the γ_2 subunit may explain the different kinetic constants for γ_1 and γ_2 homodimers. The degrees of exchanges in the α, β and γ subunits are very similar, suggesting separate but comparatively recent duplicatory diversions (Jörnvall *et al.*, 1987). The complete amino acid sequence deduced from the cDNA sequence confirms the amino acid sequences previously determined.

The class I ADH and χ-ADH genes have been located on the long arm of human chromosome 4 between q21 and q24 (Smith, 1986).

9.3.3 Metabolism of nonethanol substrates by ADH

The major physiological role of ADH is considered to be the catabolism of the great variety of naturally occurring alcohol and nonalcohol compounds of plant, animal and microbial origin. Dehydrogenation of ethanol and other alcohols, as well as many nonethanol substrates, is visualized as the main function of ADH. Alcohols arising from foodstuffs and fruits are also metabolized via the ADH pathway. ADH is capable of oxidizing a variety of primary, secondary and tertiary aliphatic alcohols and a limited number of cyclic alcohols to the corresponding aldehydes. Increase in chain length (2 to 6 carbons), unsaturation and aromatic substitution increase the effectiveness of the substrates for ADH. However, a further increase in alcohol chain length results in a decrease of substrate affinity (Pietruszko, 1979). Differences in the kinetic properties of the polymorphic forms of ADH isozymes may contribute to differences in the in vivo elimination rate of ethanol and nonethanol substrates. Since the activity of the atypical ADH_2 is several times higher than the normal enzyme at a relatively physiological pH (8.8), individuals with the atypical ADH_2 isozyme may metabolize ethanol and other aliphatic primary alcohols in a different way from individuals with a normal ADH_2 phenotype.

ADH plays an important role in retinol-retinal interconversion in visual processes. It has also been proposed that ADH is involved in the metabolism of steroids and bile acids as well as in the catabolism of short-chain alcohols responsible for food flavors (Pietruszko, 1979). Human liver ADH metabolizes digitalis compounds to their 3-keto derivatives (Frey and Vallee, 1980). Thus, ADH becomes an important biochemical link in the metabolism, pharmacology and toxicology of ethanol and cardiac glycosides and their interactions.

Isopropyl alcohol is a common source of acute intoxication in humans. It is metabolized to acetone, catalyzed primarily by ADH. Twenty per cent of isopropanol is excreted unchanged by the kidney or lungs. Acetone causes CNS depression, nausea, headache and muscle weakness. The depressive effect of higher alcohols on the CNS may be nonspecific at the level of the neuronal plasma membrane rather than interacting with specific receptors.

The metabolism of methanol and ethylene glycol by ADH also has considerable medical significance. Methanol occurs in illicit liquor and many synthetic products. Methanol is highly toxic and a small dose (about 30 ml) leads to acidosis, blindness or death. The rate-limiting step in methanol toxicity is its metabolism by ADH to the highly toxic compound formaldehyde, which is rapidly converted to the major toxin formic acid. Ethylene glycol and diethylene glycol, commonly used as antifreeze and coolant, also occur as additives in wines and other alcoholic drinks. Ethylene glycol poisoning results from its degradation by ADH to glycoaldehyde, which is further metabolized to glycolate and glyoxalic acid. These metabolites produce profound CNS toxicity and other clinical effects in humans (Winek *et al.*, 1978).

9.3.4 Aldehyde dehydrogenases

The enzymatic removal of aldehydes in humans is mediated by a number of unspecific enzymes, such as aldehyde oxidase, xanthine oxidase and ALDH. The major oxidation of acetaldehyde in the liver and other organs is catalysed by the NAD^+-dependent ALDH (aldehyde:NAD^+ oxidoreductase, EC 1.2.1.3). At least four isozymes of ALDH, coded by different gene loci, have been detected in human organs and tissues which differ in their electrophoretic mobility, kinetic properties and cellular and tissue distribution (Goedde and Agarwal, 1989). The various ALDH isozymes also differ in their molecular size, subunit structure and isoelectric point as well as in their chromosomal assignment. Hitherto, we have used a sequential designation, namely, ALDH I, II, III and IV, based on the decreasing electrophoretic mobility of the isozyme bands. However, we have now designated the ALDH isozymes according to the suggestion of Weiner (1989). Thus, the cytosolic enzyme is designated here as $ALDH_1$ or E_1, and the mitochondrial enzyme as $ALDH_2$ or E_2. On electrophoresis in starch gel, $ALDH_2$ moves faster towards the anode than $ALDH_1$, followed by $ALDH_3$ and $ALDH_4$. Both ALDH isozymes show a broad range of substrate specificity for aliphatic as well as aromatic aldehydes. While $ALDH_1$ and $ALDH_2$ show K_m values in the micromolar range with acetaldehyde and propionaldehyde, the Michaelis constants for $ALDH_3$ and $ALDH_4$ are in the millimolar range (Harada *et al.*, 1980b; Forte-McRobbie and Pietruszko, 1985; Pietruszko *et al.*, 1987). $ALDH_1$ and $ALDH_2$ isozymes are homotetramers consisting of subunits with a molecular weight of about 54 000 daltons (Greenfield and Pietruszko, 1977; Harada *et al.*, 1980b).

Subunits from both $ALDH_1$ and $ALDH_2$ isozymes are 500 residues in length (Hempel and Jörnvall, 1987). Subunit differences between the two isozymes are extensive, with only 68% positional identity (Hempel *et al.*, 1985). On the basis of partial cDNA sequencing, about 66% homology was observed between the coding regions of $ALDH_1$ and $ALDH_2$ isozymes (Hsu *et al.*, 1986). The horse and human cytosolic ALDH are more similar (91% identity) than the human cytosolic and mitochondrial isozymes (68% identity). The first 20 amino acids of the $ALDH_2$ sequence (deduced from the cDNA sequence) showed no homology to the corresponding sequence for $ALDH_1$ or $ALDH_2$ (Braun *et al.*, 1987). Moreover, an additional 15-residue peptide without a corresponding counterpart in $ALDH_1$ was found at the amino-terminal end

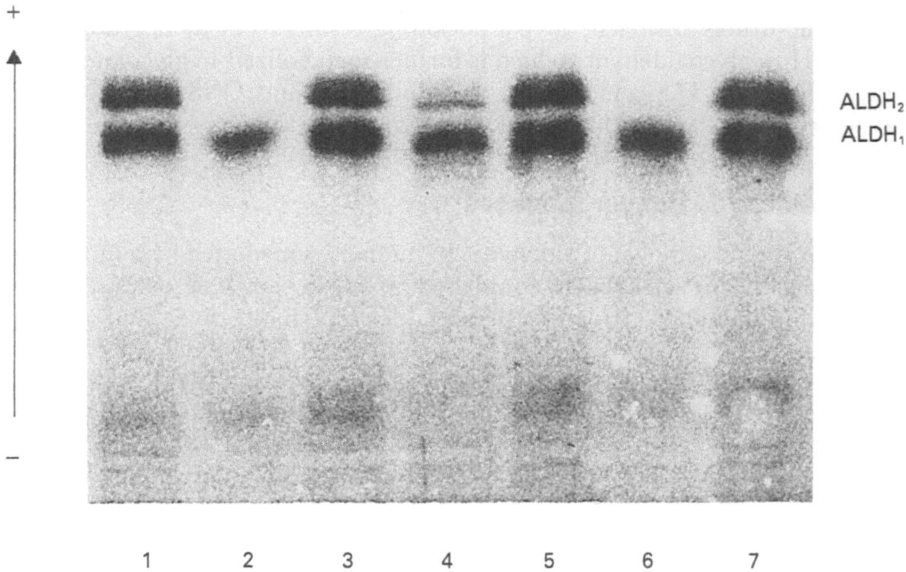

Figure 9.3 Aldehyde dehydrogenase (ALDH) isozyme pattern after isoelectric focusing of liver extracts followed by enzyme activity staining. Lanes 1, 3, 5 and 7: normal type; lanes 2, 4 and 6: deficient type. (From Agarwal and Goedde, 1990, reproduced with permission of the publisher).

of the $ALDH_2$. The whole of the 35 residue peptide seems to represent a leader peptide responsible for directing the protein import into the mitochondria.

The mitochondrial $ALDH_2$ gene is localized on chromosome 12, the cytoplasmic isozyme ($ALDH_1$) gene is located at the long arm of chromosome 9 and the $ALDH_3$ isozyme gene has been mapped at chromosome 17 (Hsu *et al.*, 1985; Santisteban *et al.*, 1985; Braun *et al.*, 1986; Goedde and Agarwal, 1986; Smith, 1986).

(a) Polymorphism of $ALDH_2$

A variation in $ALDH_2$ isozyme profile was detected in autopsy liver specimens of Japanese (Goedde *et al.*, 1979; Harada *et al.*, 1980a; Agarwal *et al.*, 1981); about 50% of the samples analysed showed no $ALDH_2$ isozyme activity band (Fig. 9.3). The cross-reacting material (CRM) isolated from livers deficient in $ALDH_2$ activity showed atypical immunochemical properties (Agarwal *et al.*, 1984). A point mutation is responsible for inactivation of the isozyme; glutamic acid at the 14th position from the C-terminus (487th position from the amino-terminus) has been substituted in the deficient isozyme with lysine (Hempel *et al.*, 1985; Hsu *et al.*, 1985).

While Oriental populations of Mongoloid origin show a varying degree of isozyme deficiency, none of the Caucasian or Negroid populations have this isozyme abnormality (Table 9.3) (Goedde *et al.*, 1983a,b, 1984, 1985, 1986; Agarwal and Goedde, 1987; Goedde and Agarwal, 1987, 1989).

Table 9.3 Distribution of the deficient phenotype of $ALDH_2$ isozyme in different populations (reproduced from Agarwal and Goedde, 1990, with permission from the publisher)

Population	*Sample size*	*$ALDH_2$ deficient (%)*
Orientals		
Ainu	80	20
Chinese		
Mongolian	198	30
Zhuang	106	25
Han	120	45
Korean (Mandschu)	209	25
Indonesians	30	39
Japanese	184	44
Koreans (South)	75	27
Phillipinos	110	13
Thais (North)	110	8
Vietnamese	138	53
South American Indians		
Atacameños (Chile)	133	43
Mapuche (Chile)	64	41
Shuara (Ecuador)	99	42
North American Indians		
Navajo (New Mexico)	56	2
Sioux (North Dakota)	90	5
Mexican Indians		
Mestizo (Mexico City)	43	4
Caucasians and blacks		
Asian Indians	50	0
Egyptians	260	0
Fangs	37	0
Germans	300	0
Hungarians	177	0
Israeli	77	0
Kenyans	23	0
Liberians	184	0
Matyo	106	0
Romai	84	0
Sudanese	40	0
Turks	65	0

Among native Indians, about 40% of the South American Indian tribes (Mapuche, Atacameños and Shuara) showed ALDH deficiency, while only a very small percentage of the North American Indians (Sioux, Navajo and Mestizo) have the isozyme deficiency. A relatively low incidence (16%) of isozyme deficiency was reported by Zeiner *et al.* (1984) in Oklahoma Indians, while no isozyme deficiency was detected in autopsy livers of American Indians of Northern New Mexico (Rex *et al.*, 1985; Bosron *et al.*, 1988). Genotyping of the $ALDH_2$ allele using allele-specific oligo-nucleotides and DNA amplification has helped to classify $ALDH_2$ variants into three groups representing normal homozygotes, heterozygotes and deficient homozygotes (Goedde *et al.*, 1989). In earlier studies using phenotyping in hair root lysates, subjects were divided into either normal or deficient types (Goedde *et al.*, 1983b, 1985, 1986). Preliminary family studies revealed a dominance of the mutant allele over the normal allele (Singh *et al.*, 1989).

9.3.5 Metabolism of other aldehydes and xenobiotics

Aliphatic and aromatic aldehydes are widespread and are present, for example, in food, smog and cigarette smoke (Schauenstein *et al.*, 1977). Biotransformation of many drugs and xenobiotics leads to the generation of aldehyde moieties. Many of these aldehydes and/or their metabolites are potentially toxic compounds. These aldehydes are primarily metabolized to alcohols by enzymatic reduction catalysed by ADH, and to the corresponding carboxylic acids by ALDH-catalysed oxidation. In particular, aromatic aldehydes, such as benzaldehyde, nitrobenzaldehyde and 2-furfuraldehyde, are preferentially catabolized by the stomach-specific $ALDH_3$. This unique class of isozymes shows extensive genetic polymorphism in liver and stomach specimens from Chinese (Yin *et al.*, 1988). However, the physiological role of $ALDH_3$ isozyme polymorphism in relation to environmental exposure to toxic aliphatic and aromatic aldehydes is not clear.

9.3.6 Ethanol- and acetaldehyde-mediated acute response to alcohol

In some individuals, ingestion of moderate amounts of alcohol leads to the so-called alcohol sensitivity symptoms, such as facial flushing, increase in heart rate, enhancement of left ventricular function, hot feeling in stomach, palpitation, tachycardia and muscle weakness. Although members of all ethnic groups seem to show biological sensitivity to alcohol associated with central nervous system and cardiovascular changes, a greater percentage of Orientals (Mongoloid ancestry) and American Indians respond to a mild dose of ethanol with marked aversive reactions as compared to Caucasians (Agarwal and Goedde, 1986, 1990; Goedde and Agarwal, 1989). The increase of plasma catecholamines apparently leads to an elevation of heart rate, dilation of peripheral vessels accompanied by a rise in blood flow in carotid arteries and increased cardiac output (Mizoi *et al.*, 1982; Adachi and Mizoi, 1983; Kupari *et al.*, 1983). A positive correlation between alcohol sensitivity and elevated blood acetaldehyde level in conjunction with $ALDH_2$ isozyme deficiency was noted in Japanese subjects given an acute dose of alcohol (Harada *et al.*, 1981; Goedde *et al.*, 1983a). Flushers with $ALDH_2$ isozyme deficiency had a significantly higher blood acetaldehyde level than nonflushers with a normal ALDH isozyme profile, while the blood ethanol concentrations were similar in both groups (Harada *et al.*, 1985).

9.3.7 Pattern of alcohol abuse and alcoholism rate in different ethnic groups

Drinking beliefs and drinking practices influenced by ethnic and cultural factors are generally considered to be mainly responsible for differences in the social drinking pattern. Epidemiological findings show that alcohol consumption varies among racial and ethnic groups. In a large-scale study comprising around 60 000 people, about 90% Caucasian, 85% Latin, 82% Japanese, 80% black, 68% Chinese and 64% Filipino subjects reported regular use of alcohol (Klatsky *et al.*, 1983). Self-reported use of three or more drinks daily was similar in Caucasians, Latins and blacks, but was significantly lower in the Asian group.

Individuals who experience unpleasant reactions after drinking alcohol may consume smaller amounts of alcohol in daily life. Generally, the Orientals flush more and drink less than the Caucasians. The role of $ALDH_2$ isozyme deficiency in affecting alcohol consumption is further supported by studies in Japan (Ohmori *et al.*, 1986). Out of 101 Japanese tested, 42% were $ALDH_2$ isozyme deficient and 58% had a normal phenotype. The subjects were divided into 'drinking habits (−)' (subjects who drank only occasionally or not at all) and 'drinking habits (+)' (who drank every day or almost every day). The subjects lived in the same city and had the same kind of occupation. When grouped according to their alcohol-drinking habits, only 19% ALDH-deficient subjects and 49% of the ALDH-normal types were in the drinking habit (+) category.

9.3.8 $ALDH_22$ isozyme abnormality and alcoholism

A significantly low incidence of $ALDH_2$ isozyme deficiency was observed in a group of alcoholics as compared to the psychiatric patients, drug dependents and healthy controls in a Japanese psychiatric hospital (Harada *et al.*, 1982; Goedde *et al.*, 1983a). In 175 alcoholics, only about

2% had the isozyme deficiency, while in drug dependents, schizophrenics and healthy controls, more than 40% had the isozyme deficiency.

Sioux and Navajo tribes and the Mestizo group showed a very low percentage of isozyme deficiency, while Native Americans from Chile and Ecuador had the deficiency to an extent comparable with that of the Japanese and Chinese. Coincidently, both the Sioux and Navajo tribes show much heavier alcohol consumption than other typical North American Indian groups (Whittaker, 1982). The study of Zeiner *et al.* (1984) also showed that $ALDH_2$ isozyme-deficient Indians drank significantly less alcohol than Indians with no isozyme deficiency; the family history of alcoholism was lower (28%) in the deficient subjects than in Indians without the isozyme deficiency (84%).

The physiological and functional implications of $ALDH_2$ polymorphism in humans are summarized as follows.

1. Genetic variations in specific enzymes involved in alcohol metabolism are important determinants of the rates of ethanol metabolism and the resulting physiological responses.
2. Slow acetaldehyde oxidation due to $ALDH_2$ isozyme abnormality leads to elevated blood acetaldehyde levels, resulting in vasodilation associated with dysphoric symptoms after drinking alcohol.
3. People differ in physiological and morphological features which influence their drinking behavior and metabolic responses to alcohol.
4. An interesting correlation exists between flushing response and alcohol-drinking habits of Japanese, Chinese and Koreans; alcohol intolerance results in lower consumption of alcohol as compared to the general population average.
5. A significantly low incidence of $ALDH_2$ isozyme deficiency is observed in Japanese alcoholics compared to psychiatric patients, drug dependents and healthy controls: only about 2% of alcoholics have the isozyme deficiency against about 40% in the control population.
6. A remarkably low deficiency of ALDH isozyme is found among North American Indians.

9.4 PUBLIC HEALTH IMPLICATIONS

In the present-day environment, human beings are constantly exposed to many chemical substances that are toxic directly or indirectly. Numerous naturally occurring alcohol and nonalcohol compounds, as well as many aliphatic and aromatic aldehydes present in alcoholic beverages and foodstuffs, are known to be potent organotoxics and carcinogens. Acetaldehyde covalently binds with a number of proteins, leading to alterations in their structure and function (Lieber, 1988). Many reports clearly indicate that genetic polymorphisms exist among enzymes which metabolize ethanol.

Interindividual differences in the catalytic and structural properties of alcohol-metabolizing enzymes can thus explain the biochemical and genetic basis of vulnerability to, or protection against, some of the toxic effects of alcohol consumption in different human populations. Individuals with variant forms of alcohol-metabolizing enzymes may be predisposed to enhanced toxic effects of higher alcohols, aldehydes and their derivatives. However, data are lacking concerning the metabolic consequences of exposure to various industrial alcohols and aldehydes in individuals with variant enzyme forms. Thus, a systematic study on interindividual and interethnic variability of enzymes of alcohol metabolism and their correlation to chemical toxicity may help in the assessment of risk and the protection of persons at high risk.

REFERENCES

Adachi, J. and Mizoi, Y. (1983) Acetaldehyde-mediated alcohol sensitivity and elevation of plasma catecholamine in man. *Jpn. J. Pharmacol.*, **33**, 531–9.

Agarwal, D. P., Eckey, R., Harada, S. and Goedde, H. W. (1984) Basis of aldehyde dehydrogenase deficiency in Orientals: immunochemical studies. *Alcohol*, **1**, 111–18.

Agarwal, D. P. and Goedde, H. W. (1986) Ethanol oxidation: ethnic variation in metabolism and response, in *Ethnic Differences in Reactions to Drugs and Xenobiotics* (eds W. Kalow, H. W. Goedde and D. P. Agarwal), Alan R. Liss, New York, pp. 99–111.

Agarwal, D. P. and Goedde, H. W. (1987) Human aldehyde dehydrogenase isozymes and alcohol sensitivity, in *Isozymes: Current Topics in Biological and Medical Research* (eds M. C. Rattazzi, J. G. Scandalios and G. S. Whitt), Vol. 16, Alan R. Liss, New York, pp. 21–48.

Agarwal, D. P. and Goedde, H. W. (1989) Enzymology of alcohol degradation, in *Alcoholism: Biochemical and Genetic Aspects* (eds H. W. Goedde and D. P. Agarwal), Pergamon Press New York, pp. 3–20.

Agarwal, D. P. and Goedde, H. W. (1990) *Alcohol Metabolism, Alcohol Intolerance and Alcoholism*, Springer-Verlag, Berlin and Heidelberg.

Agarwal, D. P., Harada, S. and Goedde, H. W. (1981) Racial differences in biological sensitivity to ethanol: the role of alcohol dehydrogenase and aldehyde dehydrogenase isozymes. *Alcoholism (NY)*, **5**, 12–16.

Azevêdo, E. S., DaSilva, M. C. B. O. and Taveres-Neto, J. (1975) Human alcohol dehydrogenase ADH_1, ADH_2 and ADH_3 loci in a mixed population of Bahia, Brazil *Ann. Hum. Genet.*, **39**, 321–7.

Bosron, W. F., Li T. K., Dafaldecker, W. P. and Vallee, B. L. (1979) Human liver π-alcohol dehydrogenase: kinetic and molecular properties. *Biochemistry*, **18**, 1101–5.

Bosron, W. F., Li, T. K. and Vallee, B. L. (1980) New molecular forms of liver alcohol dehydrogenase: isolation and characterization of $ADH_{Indianapolis}$. *Proc. Natl Acad. Sci, USA*, **77**, 5784–8.
Bosron, W. F. and Li, T. K. (1986) Genetic polymorphism of human liver alcohol and aldehyde dehydrogenase, and their relationship to alcohol metabolism and alcoholism. *Hepatology*, **6**, 502–10.
Bosron, W. F., Rex, D. K., Harden, C. A. and Li, T. K. (1988) Alcohol and aldehyde dehydrogeanse isoenzymes in Sioux North American Indians. *Alcoholism (NY)* **12**, 454–5.
Braun, T., Bober, E., Singh, S. *et al.* (1987) Evidence for a signal peptide at the amino-terminal end of human mitochondrial aldehyde dehydrogenase. *FEBS Lett.*, **215**, 233–6.
Braun, T., Grezeschik, K. H., Bober, E. *et al.* (1986) The structural gene for the mitochondrial aldehyde dehydrogenase maps to human chromosome 12. *Hum. Genet.*, **73**, 365–7.
Forte-McRobbie, C. M. and Pietruszko, R. (1985) Aldehyde dehydrogenase content and composition of human liver. *Alcohol*, **2**, 375–81.
Frey, W. A. and Vallee, B. L. (1980) Digitalis metabolism and human liver alcohol dehydrogenase. *Biochemistry*, **77**, 924–7.
Gennari, K., Wermuth, B., Muellener, D. *et al.* (1988) Genotyping of human class I alcohol dehydrogenase. Analysis of enzymatically amplified DNA with allele-specific oligonucleotides. *FEBS Lett.*, **288**, 305–9.
Goedde, H. W. and Agarwal, D. P. (1986) Aldehyde oxidation: ethnic variation in metabolism and response, in *Ethnic Differences in Reactions to Drugs and Xenobiotics* (eds W. Kalow, H. W. Goedde and D. P. Agarwal), Alan R. Liss, New York, pp. 113–38.
Goedde, H. W. and Agarwal, D. P. (1987) Polymorphism of aldehyde dehydrogenase and alcohol sensitivity. *Enzyme*, **37**, 29–44.
Goedde, H. W. and Agarwal, D. P. (1989) Acetaldehyde metabolism: genetic variation and physiological implications, in *Alcoholism: Biochemical and Genetic Aspects* (eds H. W. Goedde and D. P. Agarwal), Pergamon Press New York, pp. 21–56.
Goedde, H. W., Agarwal, D. P., Eckey, R. and Harada, S. (1985) Population genetic and family studies on aldehyde dehydrogenase deficiency and alcohol sensitivity. *Alcohol*, **2**, 283–9.
Goedde, H. W., Agarwal, D. P. and Harada, S. (1983a) The role of alcohol dehydrogenase and aldehyde dehydrogenase isozymes in alcohol metabolism, in *Isozymes: Current Topics in Biological and Medical Research* (eds M. C. Rattazzi, J. G. Scandalios and G. S. Whitt), vol. 8, Alan R. Liss, New York, pp. 175–93.
Goedde, H. W., Agarwal, D. P., Harada, S. *et al.* (1983b) Population genetic studies on aldehyde dehydrogenase isozyme deficiency and alcohol sensitivity. *Am. J. Hum. Genet.*, **35**, 769–72.
Goedde, H. W., Agarwal, D. P., Harada, S. *et al.* (1986) Aldehyde dehydrogenase polymorphism in North American, South American and Mexican Indians. *Am. J. Hum. Genet.*, **38**, 395–9.
Goedde, H. W., Benkmann, H. G., Kriese, L. *et al.* (1984) Aldehyde dehydrogenase isozyme deficiency and alcohol sensitivity in four different Chinese populations. *Hum. Hered.*, **34**, 183–6.
Goedde, H. W., Harada, S. and Agarwal, D. P., (1979) Racial differences in alcohol sensitivity: a new hypothesis. *Hum. Genet.*, **51**, 331–4.
Goedde, H. W., Singh S., Agarwal, D. P. *et al.* (1989) Genotyping of mitochondrial aldehyde dehydrogenase in blood samples using allele-specific oligonucleotides: comparison with phenotyping in hair roots. *Hum. Genet.*, **81**, 305–7.

Greenfield, N. J. and Pietruszko, R. (1977) Two aldehyde dehydrogenases from human liver: isolation via affinity chromatography and characterization of the isozymes. *Biochim. Biophys. Acta*, **483**, 35–45.

Harada, S., Agarwal, D. P. and Goedde, H. W. (1980a) Liver alcohol and aldehyde dehydrogenase in the Japanese: isozyme variation and its possible role in alcohol intoxication. *Am. J. Hum. Genet.*, **32**, 8–15.

Harada, S., Agarwal, D. P. and Goedde, H. W. (1980b) Electrophoretic and biochemical studies of human aldehyde dehydrogenase isozymes in various tissues. *Life Sci.*, **26**, 1771–80.

Harada, S., Agarwal, D. P. and Goedde, H. W. (1981) Aldehyde dehydrogenase deficiency as cause of facial flushing reaction to alcohol in Japanese. *Lancet*, **ii**, 982.

Harada, S., Agarwal, D. P. and Goedde, H. W. (1985) Aldehyde dehydrogenase polymorphism and alcohol metabolism in alcoholics. *Alcohol*, **2**, 391–2.

Harada, S., Agarwal, D. P., Goedde, H. W. *et al.* (1982) Possible protective role against alcoholism for aldehyde dehydrogenase isozyme deficiency in Japan. *Lancet*, **ii**, 827.

Hempel, J. and Jörnvall, H. (1987) Functional topology of aldehyde dehydrogenase structures, in *Enzymology and Molecular Biology of Carbonyl Metabolism: Aldehyde Dehydrogenase, Aldo-Keto Reductase and Alcohol Dehydrogenase* (eds H. Weiner and T. G. Flynn), Alan R. Liss, New York, pp. 1–14.

Hempel, J., Kaiser, R. and Jörnvall, H. (1985) Mitochondrial aldehyde dehydrogenase from human liver: primary structure, differences in relation to the cytosolic enzyme and functional correlations. *Eur. J. Biochem.*, **153**, 13–28.

Hsu, L. C., Tani, K., Fujiyoshi, T. *et al.* (1985) Cloning of cDNAs for human aldehyde dehydrogenases 1 and 2. *Proc. Natl Acad. Sci. USA*, **82**, 3771–5.

Hsu, L. C., Yoshida, A. and Mohandas, T. (1986) Chromosomal assignment of the genes for human aldehyde dehydrogenase–1 and aldehyde dehydrogenase–2. *Am. J. Hum. Genet.*, **38**, 641–8.

Ikuta, T., Shibuya, A. and Yoshida, A. (1988) Direct determination of usual and atypical alleles on the human aldehyde dehydrogenase–2 ($ALDH_2$). *Biochem. Genet.*, **26**, 519–25.

Jörnvall, H., Hempel, J. and Vallee, B. (1987) Structures of human alcohol and aldehyde dehydrogenases. *Enzyme*, **37**, 5–18.

Juliá, P., Boleda, M. D., Farrés, J. and Parés, X. (1987) Mammalian alcohol dehydrogenase: characteristics of class III isozymes. *Alcohol Alcoholism*, (Suppl. 1), 169–74.

Klatsky, A. L., Siegelaub, A. B., Landy, C. and Friedman, G. D. (1983) Racial patterns of alcoholic beverage use. *Alcoholism (NY)*, **7**, 372–7.

Kupari, M., Eriksson, C. J. P., Heikkilä, J. and Ylikahri, R. (1983) Alcohol and the heart. Intense hemodynamic changes associated with alcohol flush in Orientals. *Acta Med. Scand.*, **213**, 91–8.

Lieber, C. S., (1988) Metabolic effects of acetaldehyde. *Biochem. Soc. Trans*, **16**, 241–7.

Martin, N. G., Perl, J., Oekeshott, J. G. *et al.* (1985) A twin study of ethanol metabolism. *Behav. Genet.*, **15**, 93–109.

Mizoi, Y., Adachi, J., Kogame, M. and Fukunaga T. (1982) Polymorphism of aldehyde dehydrogenase and catecholamine metabolism, in *Enzymology of Carbonyl Metabolism: Aldehyde Dehydrogenase and Aldo/Keto Reductase* (eds H. Weiner and B. Wermuth), Alan R. Liss, New York, pp. 363–77.

Ohmori, T., Koyama, T., Chen, C. *et al.* (1986) The role of aldehyde dehydrogenase isozyme variance in alcohol sensitivity, drinking habits formation and the development of alcoholism in Japan, Taiwan and the Philippines. *Prog. Neuropsychopharmacol. Biol. Psychiatry*, **10**, 229–35.

Parés, X. and Vallee, B. L. (1981) New human liver alcohol dehydrogenase forms with unique kinetic characteristics. *Biochem. Biophys. Res. Comm.*, **98**, 122–30.

Pietruszko, R. (1979) Nonethanol substrates of alcohol dehydrogenase, in *Biochemistry and Pharmacology of Ethanol* (eds E. Majchrowicz and E. P. Noble), Plenum Press, New York, pp. 87–106.

Pietruszko, R., Ryzlak, M. T. and Forte-McRobbie, C. M. (1987) Multiplicity and identity of human aldehyde dehydrogenases. *Alcohol Alcoholism*, (Suppl. 1), 175–9.

Rex, D. K., Bosron, W. F., Smialek, J. E. and Li, T. K. (1985) Alcohol and aldehyde dehydrogenase isoenzymes in North American Indians. *Alcoholism (NY)*, **9**, 147–52.

Santisteban, I., Povey, S., West, L. F. *et al.* (1985) Chromosome assignment, biochemical and immunological studies on a human aldehyde dehydrogenase, ALDH3. *Ann. Hum. Genet.*, **49**, 87–100.

Schauenstein, E., Esterbauer, H. and Zollner H. (1977) *Aldehydes in Biological Systems: Their Natural Occurrence and Biological Activities*, Pion, London, pp. 1–205.

Singh, S., Fritze, G., Fang, B. *et al.* (1989) Inheritance of mitochondrial aldehyde dehydrogenase: genotyping in Chinese, Japanese and South Korean families reveals a dominance of the mutant allele. *Hum. Genet.*, **83**, 119–21.

Smith, M. (1986) Genetics of human alcohol and aldehyde dehydrogenases, in *Advances in Human Genetics* (eds H. Harris and K. Hirschhorn), Vol. 15, Plenum Press, New York, pp. 249–90.

Smith, M., Hopkinson, D. A. and Harris, H. (1972) Alcohol dehydrogenase isozyme in adult human stomach and liver: evidence for activity of the ADH_3 locus. *Ann. Hum. Genet.*, **35**, 243–53.

Smith, M., Hopkinson, D. A. and Harris, H. (1973) Studies on the subunit structure and molecular size of the human alcohol dehydrogenase isozymes determined by the different loci ADH_1, ADH_2, ADH_3. *Ann. Hum. Genet.*, **36**, 401–14.

Teng. Y-S., Jehan, S. and Lie-Injo, L. E. (1979) Human alcohol dehydrogenase ADH_2 and ADH_3 polymorphism in ethnic Chinese and Indians of West Malaysia. *Hum. Genet.*, **53**, 87–90.

Vesell, E. S., Page, J. G. and Passananti, G. T. (1971) Genetic and environmental factors affecting ethanol metabolism in man. *Clin. Pharmacol. Ther.*, **12**, 92–201.

von Wartburg, J. P., (1989) Pharmacokinetics of alcohol, in *Human Metabolism of Alcohol* (eds K. E. Crow and R. D. Batt), Vol. 1, CRC Press, Boca Raton, pp. 9–22.

Weiner, H. (1989) Letter to the editor. *Alcoholism (NY)*, **13**, 425.

Whittaker, J. O. (1982) Alcohol and the Standing Rock Sioux tribe. A twenty-year follow-up study. *J. Stud. Alcohol*, **43**, 191–200.

Winek, C. L., Shingleton, D. P. and Shanor, S. P. (1978) Ethylene and diethylene glycol toxicity. *Clin. Toxicol.*, **13**, 297–324.

Xu, Y., Carr, L. G., Bosron, W. F. *et al.* (1988) Genotyping of human alcohol dehydrogenases at the ADH2 and ADH3 loci following DNA sequence amplifications. *Genomics*, **2**, 209–14.

Yin, S. J., Cheng T. C., Chang, C. P. *et al.* (1988) Human stomach alcohol and aldehyde dehydrogenases (ALDH): a genetic model proposed for ALDH III isozymes. *Biochem. Genet.*, **26**, 343–60.

Zeiner, A. R., Girardot, J. M., Nichols, N. and Jones-Saumty, D. (1984) ALDH I isozyme deficiency among North American Indians. *Alcoholism (NY)*, **8**, 129.

10 Detection of unsuspected, non-fava-bean-sensitive G–6–PD deficiency in semiconductor fabrication workers in Israel

D. T. Teitelbaum

Medical Toxicology Partnership, Denver, Colorado, USA

and B. Raikhlin

Israel National Poison Information Center, Haifa, Israel

10.1 SUMMARY

Glucose 6-phosphate dehydrogenase (G–6-PD) deficiency is the most common genetic disorder of human erythrocyte metabolism; it has been observed in many communities in Israel. Families in which favism has been diagnosed are aware that some drugs and chemicals may produce haemolytic crises. However, those with G–6-PD deficiency not sensitive to the fava bean may be unaware of the problem until they are exposed to a haemolysin and undergo a crisis. The Israel National Poison Information Center became concerned that a simple inquiry concerning favism might not detect persons likely to undergo haemolysis in acute poisoning.

In a study of 60 individuals employed in a semiconductor fabrication facility with potential exposure to chemicals which may produce haemolysis in G–6-PD-deficient persons, four previously undiagnosed and unsuspected cases of G–6-PD deficiency were noted. This suggests that the incidence of unsuspected non-fava-bean-sensitive G–6-PD deficiency may be as high as 11% in males of Sephardi Jewish families and in the Israeli Arab population. If this is the true incidence, pre-employment screening for G–6-PD deficiency would be appropriate in those industries where there is potential for exposure to chemicals which produce haemolysis. It appears that the standard practice of inquiry concerning favism in the immediate family may not be sufficient.

10.2 G–6-PD DEFICIENCY

Erythrocyte G–6-PD deficiency is believed to be the most common genetically-determined abnormality of red cell metabolism in the world. Vulliamy *et al.* (1988) suggest that there are more than 300 separate, identified G–6-PD variants. More than 400 million people are affected by some variety of G–6-PD deficiency. Electrophoretically Type A cases are most common in Africa. Elsewhere Type B cases predominate. G–6-PD-deficient heterozygotes gain some protection from malaria, which may explain the persistence of this widespread polymorphic genetic trait.

10.3 DIVERSITY OF MANIFESTATIONS OF G–6-PD DEFICIENCY

Many individual variant electrophoretic patterns of G–6-PD have been observed. Clinically, there is a broad spectrum of effect from G–6-PD deficiency. Some patients have congenital non-spherocytic anaemia; some haemolyse with relatively minimal stress; some demonstrate favism and some are entirely asymptomatic unless they are exposed to a drug or chemical which produces oxidative stress. A comprehensive review of the similarities and differences among the variants is beyond the scope of this paper.

The diversity of electrophoretic patterns suggests that multiple point mutations are responsible for the variegated picture of G–6-PD deficiency. Wherever in the world G–6-PD deficiency has been sought, it has been found. Although it is tempting to use G–6-PD variants to trace particular ethnic strains in larger populations (Wijsman, 1984) this may not be a valid approach. Since new mutations occur frequently, and the G–6-PD gene may not be stable, there may be insufficient information to form conclusions concerning population genetics.

10.4 FUNCTION OF G–6-PD

G–6-PD catalyses the first step in the hexose monophosphate pathway, or HMP shunt, of aerobic glycolysis in the erythrocyte. The HMP shunt is the second of the two principal pathways of glucose metabolism in the red cell. G–6-P is oxidized to 6-phosphogluconolactone by NADP, which is reduced to NADPH. The 6-phosphogluconolactone hydrolyses spontaneously to 6-phosphogluconate, the substrate for 6-phosphogluconate dehydrogenase and NADP. Further steps in this pathway result in the reduction of NADP to NADPH and the isomerization of the ribulose–5-phosphate to ribose–5-phosphate. Eventually, metabolites in this series of enzymic reactions enter the Emden-Myerhoff (E-M) pathway, to continue through the major pathway of glucose metabolism.

10.5 CONSEQUENCES OF G–6-PD DEFICIENCY

In normal, unstressed erythrocytes (Beutler, 1983) only 10% of the G–6-P enters the HMP pathway. Ninety per cent is metabolized anaerobically in the E-M pathway. The rate of metabolism depends upon the rate of NADPH oxidation on the one hand and on the activity of G–6-PD (in deficiency, for instance) on the other. Individuals with G–6-PD deficiency appear unable to provide NADPH at a rate sufficient to maintain glutathione in a reduced form during exposure to certain drugs and chemicals whose metabolism is NADPH dependent.

Decreased activity of G–6-PD in the mature erythrocytes of deficient individuals causes a shift in metabolic balance between the HMP and E-M pathways. Ultimately, the imbalance leads to denaturation of haemoglobin and red cell lysis.

Whether the deficiency of G–6-PD is the result of decreased production of enzyme molecules, the formation of molecules of less than normal activity or less than normal stability of the enzyme is the subject of much debate. Whichever mechanism may actually be the cause of dysfunctional G–6-PD, the net result in the erythrocyte under oxidative stress is similar. Haemoglobin is denatured, the characteristic Heinz bodies are formed, the membrane of the erythrocyte becomes unstable and the cell lyses.

Patients in haemolytic crisis due to G–6-PD deficiency may become very ill and die. If appropriate therapy is instituted promptly, renal failure may be prevented, which significantly improves the prognosis. It is of considerable clinical value to know which individuals are likely to undergo haemolytic crisis as a result of G–6-PD deficiency. It is also useful to know which substances can produce oxidative stress. In a Poison Information Centre, both elements of the data are used to set out a therapeutic protocol for poisoning with a material likely to produce haemolysis in G–6-PD-deficient individuals. Some commonly encountered drugs and chemicals which produce haemolysis in G–6-PD-deficient patients are shown in Table 10.1

Table 10.1 Drugs shown to cause haemolysis in G–6-PD deficiency

Acetanilid	Phenylhydrazine
Aspirin	Primaquine
Amyl nitrite	Salicylic acid
Butyl nitrite	Sulfacetamide
Coal tar derivatives	Sulfanilamide
Methylene blue	Sulfapyridine
Nalidixic acid	Synthetic benzenes
Naphthalene	Thiazolesulfone
Nitrofurantoin	Toluidine blue
Pamaquine	Trinitrotoluene
Pentaquine	

10.6 G–6-PD DEFICIENCY IN ISRAEL

In Israel, many individuals are of Middle Eastern, Levantine and North African origin. Jews from Asia Minor and East Africa form a significant portion of the population. Jewish families from Western Europe, Russia, the Balkans and the New World contribute to a highly varied ethnic picture. Arabs, Druze, Armenians, Circassians and many other ethnic groups are also part of the Israeli population.

On the basis of studies performed by Szeinberg (1959) and others in the 1950s and 1960s, it is believed that G–6-PD deficiency in Israel is found predominantly in Sephardi Jewish families. However, studies in other countries have demonstrated G–6-PD deficiency in Ashkenazi Jewish families. Russia, Bulgaria, Italy, France, the USA (Luzzatto, 1974) are among the countries in which Jewish families with G–6-PD deficiency have been reported. Electrophoretically, these appear to be B− variants, which are highly sensitive to oxidative-stress-induced haemolysis.

10.7 G–6-PD DEFICIENCY AND FAVISM

Some of the G–6-PD-deficient variants which have been reported in Greece and elsewhere are not sensitive to fava beans. Although only G–6-PD-deficient individuals develop favism, not all deficient persons have the disorder. It appears that additional genetic factors beyond the G–6-PD status are needed for favism to become manifest (Kattamis *et al.* 1969). Thus, some individuals with Mediterranean type B− G–6-PD deficiency may remain unaware of their deficiency. In addition, some patients have never eaten fava beans, so that their sensitivity to Vicia faba would not be a useful diagnostic criterion.

Since some G–6-PD patients will not manifest favism, but may still haemolyse when stressed with various drugs or chemicals, the screening technique used to to detect G–6-PD deficiency at the Israel National Poison Information Center may not suffice. To inquire, in cases of poisoning, whether the patient and his family eat 'ful', the fava bean, without difficulty and to examine the ethnicity as indicated by name and family history may not be enough to assure reliable detection of patients at risk of haemolysis due to G–6-PD deficiency and oxidative stress. G–6-PD erythrocyte screening is not routinely performed at birth, in pre-school or school examinations, at entrance to the army or in pre-employment examinations in Israel. Therefore, there is a void in the information which is available.

10.8 ETHNIC BACKGROUND OF THE ISRAELI WORK FORCE

In many industries in Israel, persons of Sephardi background constitute a high percentage of the work force. Because G–6-PD deficiency has been associated with Sephardi communities from Iraq, Iran, Turkey and various North African countries, it has become apparent at the Israel National Poison Information Center that special attention should be paid to persons of Sephardi background who are exposed to various chemicals known to produce haemolysis. However, Israeli society is becoming increasingly integrated. Many marriages cross traditional communal lines. Many of the new-generation Israelis are of mixed Sephardi and Ashkenazi parentage. Since G–6-PD deficiency is sex-linked and may not be expressed in the heterozygous female, families with male children of mixed parentage may be surprised to discover the presence of G–6-PD deficiency, with or without favism, in youngsters with Ashkenazi surnames.

Development of 'high-technology' industry in Israel is a major economic priority. A number of the work force in such industries are exposed to organic chemicals, some of which are capable of producing oxidative stress and haemolysis in G–6-PD-deficient individuals (Table 10.2). We have therefore studied workers in the semiconductor industry, a high-technology industry which uses chemicals.

Table 10.2 Semiconductor manufacturing chemicals which may induce haemolysis in G–6-PD deficient workers

Aromatic naphtha	Nitro-organics
Alkyl benzenes	Quinones
Benzene	Quinolones
Cresols	Toluene
Naphthalene	Xylenes

10.9 MATERIALS AND METHODS

In order to obtain volunteers from a relevant industry, inquiries were made of semiconductor manufacturing facilities in Israel. There are approximately 2000 workers in Israel involved in fabrication of semiconductors. Although there was a general recognition of the importance of the study which we proposed to do, labour/management issues, security issues, medical policies and other impediments were encountered which substantially delayed completion of the study.

Although other 'high-technology' industries such as petrochemicals refining and formulating, munitions production, printing, publishing and food sciences might also have been suitable for the study it proved difficult to find an employer willing to participate in the study in those industries.

We preferred to study the semiconductor industry because chemical exposures in the industry have been very well characterized in the USA and elsewhere, and because the Israeli semiconductor industry has very proactive and dynamic health and safety personnel who, we believed, would understand the value of the study. Fortunately, National Semiconductor Corporation, an American company with a major manufacturing facility in Israel, agreed to study of volunteers at its plant in Migdal HaEmek.

The area surrounding Migdal HaEmek is heavily populated by Jews of Sephardi background. Families from Morocco, Libya, Algeria, Tunisia, Iran and Iraq have settled there, and constitute a very significant portion of the work force at the National Semiconductor Fabrication Facility. This provided an ideal population in which to examine the G–6-PD status.

Sixty individuals volunteered to participate in this study: 52 males and 8 females. The age range was 20 to 45 years. Thirty-four subjects were of Mediterranean background, 20 were of European background and 6 were of mixed background.

Each subject was interviewed prior to the collection of a blood sample. The purpose of the study was explained and consent obtained. The worker was asked about his or her background and ethnicity for two generations on maternal and paternal sides. Each was asked whether he or she was aware of G–6-PD deficiency in him/herself or in any close relative. All workers were asked if they ate 'ful' (fava beans). Several workers were aware of G–6-PD deficiencies in close relatives, none had any personal awareness of G–6-PD deficiency.

A 5-ml blood sample from each volunteer was drawn into a Vacutainer® containing EDTA. The blood was thoroughly mixed by inversion. The sample was cooled and analysed within 12 hours of collection for haemoglobin content and G–6-PD activity. The G–6-PD activity analyses were performed in the blood bank of Rambam Medical Center in Haifa by a modification of the method of Loder and Gruchy (1965).

As soon as the results had been tabulated they were returned to each of the individuals who had volunteered to participate in the study.

10.10 ASSAY FOR G–6-PD

Among the 60 workers whose blood was analysed, four individuals were found to have no detectable G–6-PD activity. This finding is consistent with a Mediterranean-type G–6-PD deficiency. All were males. One individual, age 34, was of Iraqi background; one, aged 30, was of Yemenite background; one, age 28, was of mixed Moroccan and Iraqi background and one, age 32, was of Israeli Arab background. None of these individuals was aware of G–6-PD deficiency in himself or in his family. None had at any time suffered any known reaction to fava beans or to any other known cause of haemolysis in G–6-PD-deficient individuals.

All haemoglobin determinations were normal. All other individuals were found to have normal G–6-PD activity(between 3.2 and 6.5 IU/g haemoglobin).

10.11 DISCUSSION

The finding of an unsuspected G–6-PD deficiency in 4 out of 52 males, of whom 34 were of Sephardi or Mediterranean background in whom favism had not occurred, is striking. This is an incidence of 11% non-fava-bean-sensitive G–6-PD deficiency in this group of Mediterranean-background individuals.

Although previous studies have reported G–6-PD deficiency in various families in Israel, many of these studies are more than 20 years old (Ramot *et al.*, 1969). They were performed at a time when ethnic identities were more apparent than they are in Israel today. Because of the high rate of intermarriage among various ethnic groups in Israel, the occurrence of G–6-PD deficiency in unsuspected populations is to be anticipated. As further intermixing of populations take place, it is to be expected that G–6-PD deficiency will be noted in families with Ashkenazi surnames and without Sephardi identity. In addition, there are those families which may be expected to demonstrate this trait as a result of the strains of G–6-PD deficiency known to have occurred in Russia (Shatskaya *et al.*, 1976), Bulgaria, France, Hungary, Italy, Germany, North America and other countries where Jews have lived. What is particularly interesting in this population is the finding of G–6-PD deficiency in the absence of favism in all of the cases in our study.

The semiconductor fabrication industry is chemically intensive. Among the materials which are regularly used, there are substances which have the potential to produce haemolysis in G–6-PD-deficient individuals on an acute and on a low-grade chronic basis (Table 10.2). Many nitro-organics are used in various processes in this industry as photoactivators. In addition, quinolones, quinones and other benzene derivatives, which are capable of producing abnormal redox states, are used in photoresist chemicals which are essential to the semiconductor fabrication process (Teitelbaum, 1986). It was appropriate to study the incidence of this G–6-PD deficiency in the semiconductor industry as a model for other chemically-intensive industries.

Although the semiconductor industry is at the forefront of occupational safety and health in Israel, the possibility of exposure to highly toxic materials at trace concentrations or low levels always exists. The potential for sudden massive exposure is also real. To date, we do not have any evidence of illness in the semiconductor industry in Israel which might be traceable to G–6-PD deficiency. However, no studies of this problem have been done.

Although the Israel National Poison Information Center makes it a practice to ask each person who is exposed to a material known to produce haemolysis if they are G–6-PD deficient on the basis of the occurrence of favism in themselves or first-order relatives, this question may be invalid in the light of the results of this study. Since the G–6-PD deficiency detected in this group is of the type in which favism is not a factor, the approach which we currently use would not be sufficient to warn of impending acute haemolysis in G–6-PD-deficient individuals who are not also sensitive to fava beans.

The pressure of time and limited resources available for the initial study did not permit electrophoretic determination of the G–6-PD pattern. Nor was it possible to study random callers to the Israel National Poison Information Center to determine G–6-PD status. In a larger study it would be appropriate not only to screen for G–6-PD activity in industry and in the callers to the center, but also to undertake electrophoretic analysis of the positive cases in order to determine the type of disorder which is present.

On the basis of this study, it is believed that a larger and more comprehensive study of populations in Israel for unsuspected G–6-PD deficiency is indicated. The study should include determination of G–6-PD activity, electrophoretic patterns and complete blood counts in order to assess the prevalence of G–6-PD deficiency, its type and its consequences in the Israeli population.

10.12 SUMMARY AND CONCLUSIONS

In a study of 60 individuals employed in a semiconductor fabrication facility in Israel with potential exposure to chemicals which may produce haemolysis in G–6-PD-deficient persons, four previously undiagnosed and unsuspected cases of G–6-PD deficiency of a severe type were noted. No evidence of clinical disease was obtained. These results suggest that the incidence of unsuspected non-fava-bean-sensitive G–6-PD deficiency may be as high as 11% in males of Sephardi Jewish families and in the Israeli Arab population. No cases of deficiency were detected in females, although one would not have expected to find more than two cases, according to the Hardy-Weinberg equilibrium principle in a stable population.

If this is the true incidence, then pre-employment screening for G–6-PD deficiency would be appropriate in those industries in which there is a potential for exposure to chemicals which produce haemolysis. It appears that the standard practice of inquiry concerning favism in the immediate family may not be sufficient to provide a clue to the worker's G–6-PD status.

ACKNOWLEDGEMENTS

The authors wish to thank Mr Irad Shaham, Director of Health and Safety, National Semiconductor Corporation, Migdal HaEmek, Israel, and the personnel of National Semiconductor Corporation for their participation in this study. Thanks are also due to Ms Sarah Streichmann and the staff of the haematology laboratory of the Blood Bank at the Rambam Medical Center in Haifa for their assistance in performing the G–6-PD analyses.

REFERENCES

Beutler, E. (1983) Glucose–6-phosphate dehydrogenase deficiency, in *The Metabolic Basis of Inherited Disease* (eds J. B. Stanbury *et al.*), 5th ed, McGraw-Hill, New York, pp. 1629–53.

Kattamis, C. A., Kyriazakou, M. and Chaidas, S. (1969) Favism: clinical and biochemical data. *J. Med. Genet.*, **6** (1), 34–41.

Loder, P. B. and De Gruchy, G. C. (1965) Red-cell enzymes and co-enzymes in non-spherocytic congenital haemolytic anaemias. *Br. J. Haematol.*, **2**, 21–31.

Luzzatto, L. (1974) Genetic heterogeneity and pathophysiology of G6PD deficiency. *Br. J. Haematol.*, **28**(2), 151–5.

Ramot, B., Ben-Bassat, I. and Shchory, M. (1969) New glucose–6-phosphate dehydrogenase variants observed in Israel and their association with congenital nonspherocytic hemolytic disease. *J. Lab. Clin. Med.*, **74**(6), 895–901.

Shatskaya, T. L., Krasnopolskaya, K. D. and Idelson, L. J. (1976) Mutant forms of erythrocyte glucose–6-phosphate dehydrogenase in Ashkenazi Jews. Description of two new variants: G6PD Kirovograd and G6PD Zhitomir. *Hum. Genet.*, **33**(2), 175–8.

Szeinberg, A., Pras, M., Sheba, C. *et al.* (1959) The hemolytic effects of various sulfonamides on subjects with a deficiency of glucose–6-phosphate dehydrogenase of erythrocytes. *Isr. J. Med. Sci.*, **18**, 176.

Teitelbaum, D. T. (1986) Photoactive chemicals used in photoresist systems. *State Art Rev. Occup. Med.*, **1**(1), 59–68.

Vulliamy, T. J., D'Urso, M., Battistuzzi, G. *et al.* (1988) Diverse point mutations in the human glucose–6-phosphate dehydrogenase gene cause enzyme deficiency and mild or severe hemolytic anemia. *Proc. Natl. Acad. Sci. USA*, **85**(14), 5171–5.

Wijsman, E. M. (1984) Techniques for estimating genetic admixture and applications to the problem of the origin of the Icelanders and the Ashkenazi Jews. *Hum. Genet.*, **67**(4), 441–8.

PART FIVE

Toxicogenetic Evidence

11 Prevalence of important polymorphisms in Poland

J. Z. Hanke

Institute of Occupational Medicine, Lodz, Poland

11.1 SUMMARY

The prevalence of major polymorphisms has been the subject of about 20 studies in Poland. The results are for the first time reviewed here in English. The studies focused on glucose–6-phosphate dehydrogenase (G–6-PD), alpha$_1$-antitrypsin, methaemoglobin reductase and *N*-acetyltransferase. The analytical methods must be validated to allow comparison between different studies. The results may be influenced by occupational exposures. Also, such genetic markers may be important determinants for the development of cancer. Evidence in this area must be used to protect hypersusceptible individuals.

11.2 INTRODUCTION

Genetic factors are likely to enhance the risk of exposed individuals developing environmentally-induced effects. In particular, some homozygous conditions may result in a greatly increased risk, but little is known about risks associated with the heterozygous state. A major research challenge is the development of accurate, quick and inexpensive tests to identify heterozygous carriers of traits. Because the heterozygotes of many states have a high frequency in the population, such information would be of wide clinical and epidemiological significance.

In the period 1968 to 1988, more than 20 publications were prepared concerning the prevalence of important polymorphisms in Poland. All these reports were published in Polish and, therefore, are almost inaccessible to English-speaking scientists. For this reason, a short review of findings and conclusions seems justified. The material covers four major areas: G–6-PD, alpha$_1$-antitrypsin, methaemoglobin reductase and *N*-acetyltransferase.

The following scheme has been adopted. First, results are given concerning the frequency of so-called deficits or biochemical abnormalities as ascertained in the Polish population studies. Next, observations are reviewed concerning the effect of chemical substances (industrial exposure) or particular pathological conditions (i.e. diseases) on the examined proteins. While analysing such conditions, it is sometimes difficult to distinguish between the effect and the cause. Special attention has been paid to the relationship between the biochemical parameters (deficit or biochemical alteration) and cancer, as biochemical indicators of carcinogenic risk are of great interest. The last item of the review is devoted to the methodological problems of mass screening.

11.3 G–6-PD

The interest of physicians in G–6-PD originates from the finding by Carson *et al.* (1956) that enzyme deficiency caused haemolytic reactions.

The first examinations concerning G–6-PD deficiency in the Polish population were carried out by our team in 1968 using the G–6-PD spot test produced by Calbiochem. The test is based on the reduction of tetrazole dye (3,4,5-dimethylthiazole 1,2-diphenyl-tetrazole bromide) by $NADPH^+$. The reduced form of NADP is generated in the presence of G–6-PD, which catalyses the conversion of glucose–6-phosphate to 6-phosphogluconic acid. Phenazine methylsulfate (Fairbanks and Beutler, 1962) acts in this system as a hydrogen donor.

We examined 529 people (250 females and 279 males) between 1 day and 80 years of age (Wierzbowska and Hanke, 1968). No case of G–6-PD deficiency was found. As negative control we used erythrocytes from members of families where favism had occurred. A more comprehensive study (Hanke *et al.*, 1968) was performed on a group of 1000 people, 200 women and 800 men, between 15 and 50 years of age. All individuals originated from Lodz or the region around this city. We used three different analytical methods (Brewer *et al.*, 1969; Berger, 1963; Sass *et al.*, 1966). The method of Berger (1963) is a colorimetric one and is based on the generation of reduced NADP formed under the influence of G–6-PD. The $NADPH^+$ reduces brilliant cresyl blue to a colourless compound. The rate of disappearance of the colour is thus proportional to the enzyme activity. The method of Brewer *et al.* (1960) is based on the following principle. Methaemoglobin generated by the action of sodium nitrate in the presence of methylene blue is spontaneously reduced; this reaction is caused by G–6-PD in the red blood cell. The rate of methaemoglobin disappearance thus depends on the G–6-PD activity. In the method of Sass *et al.* (1966), the rate of methylene blue disappearance is utilized as the measure of G–6-PD activity; the dye is the only reagent necessary for the measurement. Only two individuals exhibited definite G–6-PD deficiency; both were identified by the procedure according to Brewer *et al.* (1960). On the basis of our results (Hanke *et al.*, 1968; Linkowska-Pinkiewicz, 1979), we concluded that the frequency of G–6-PD deficiency in the Polish population seems to be quite similar to that in other countries in Middle Europe, that is about 0.1%.

More recently, we expanded out study of G–6-PD deficiency to nonoccupationally-exposed individuals in the Polish population (Hanke *et al.*, 1975). This expansion was promoted by the finding that changes in erythrocytic enzyme activities occur as a result of chemical exposures. We performed 400 measurements of G–6-PD activity by the method of Sass *et al.* (1966). This group comprised 349 men and 51 women not occupationally exposed to noxious agents. The statistical analysis documented a highly significant right-sided asymmetry. Age was not a parameter that influenced the enzyme activity, but the G–6-PD activity in the spring–summer period was slightly higher than in the autumn–winter period. The dispersion in the female group was broader than in the male group, but the difference between the mean values was not statistically significant. The probability of values 50% lower than the geometrical mean was about 1.4% including real representatives of G–6-PD deficiency carriers.

In 1979, 132 foreign students in Poland were examined (Linkowska-Pinkiewicz, 1979). This group comprised 108 men and 26 women, originating from Africa, Asian, USA and from the Mediterranean region of Europe. The G–6-PD activity was measured with a classical procedure (Löhr and Waller, 1970) and with two screening methods (Brewer *et al.*, 1960; Aldrich *et al.*, 1974). Comparison between the methods established their usefulness for mass population studies (Linkowska-Pinkiewicz, 1980). Eleven individuals with G–6-PD deficiency were identified: eight people with a considerably lower enzyme activity and three with a partial lack of enzyme activity. Six males formed a separate subpopulation characterized by a clearly diminished enzyme activity. Statistical analysis of the distribution of the rest of the values obtained indicated a normal distribution. Thus, we can assume that, in Poland, some people belong to the risk group characterized by a frequency of G–6-PD deficiency much higher than in the general population.

A more comprehensive study is being performed to gather more detailed data on G–6-PD deficiency in Poland (see below).

11.3.1 Changes in G–6-PD activity due to external agents and cancer

In considering the significance of these observations, it is important to recognize that the measured level of enzyme activity is a complex parameter. Exogenous factors, such as specific inhibitors, cofactors and metal ions, may affect the activity observed (Hanke, 1968; Dabrowski *et al.*, 1980). Also some diseases and treatments could potentially influence the G–6-PD activity in red blood cells; a good example is psoriasis (Kaczorowska-Hanke *et al.*, 1971). Also, some reports indicate a lower incidence of cancer in populations with higher frequencies of G–6-PD deficiency (for reviews see Hanke, 1984).

In 1980, a Polish research group (Fleiger *et al.*, 1980) observed a significantly higher G–6-PD activity in red blood cells of patients with malignant and nonmalignant maxillary tumours than in those of a control group. The same tendency was observed in rats with transplantable Morris hepatoma. The G–6-PD activity was measured using the Glook and McLean method (Tysper and Karon, 1971). The ultimate state of this experimental disease was associated with an increase of erythrocytic G–6-PD activity with a parallel decrease of glutathione (GSH). Also, our own observations showed that after surgical removal of cancer tissue, the erythrocytic G–6-PD activity drops. Thus G–6-PD activity in the red blood cells may be considered an early symptom of cancer or a tumour marker.

11.3.2 Methodological problems

The usefulness of five methods for screening of G–6-PD deficiency was assessed using data on 52 students (43 men and 9 women) from tropical Africa. The maximum number of persons with decreased G–6-PD activity (n = 11) was found using the cyanmethaemoglobin elution test (Beutler and Balude, 1963; Kleinhauer and Betke, 1963), which seems to be concordant with literature indicating that this method is the most sensitive. The highest correlation was found between the results of the cyanmethaemoglobin elution test, the Heinz bodies test (Beutler *et al.*, 1955) and the method of Löhr and Waller (1970). The smallest correlation index was found between the results of the glutathione instability test (Beutler, 1957) and the method of Löhr and Waller (1970). In three cases the results of the glutathione instability test indicated G–6-PD deficiency that was not confirmed by other methods. In three cases, G–6-PD deficiency was found with other methods but not with the glutathione instability test. Taking into account economic factors, the methaemoglobin reduction test (Brewer *et al.*, 1960) is most suitable for mass screening.

11.4 ALPHA$_1$-ANTITRYPSIN

In 1963, Laurell and Eriksson reported a relationship between the deficiency of alpha$_1$-antitrypsin (alpha$_1$-AT) and the incidence of chronic obstructive lung diseases. Some alpha$_1$-AT deficiency phenotypes may also favour the development of some hepatopathies exhibited generally in children, but also in adults (Triger *et al.*, 1976).

11.4.1. Frequency of alpha$_1$-AT deficiency in the Polish population

The first results concerning the frequency of different alpha$_1$-AT phenotypes (Pi-phenotypes) and alpha$_1$-AT deficiency in Poland were reported in 1974 by Opolska. Using gel electrophoresis, the author examined 3500 serum samples from southern Poland and established the following frequencies of particular phenotypes: Pi^M = 98.17%, Pi^S = 1.56%, Pi^F = 6.07%, Pi^Z = 0.14%, and Pi^I = 0.01%.

In 1979, Pilacik and Kotulska-Hofmokl published results of a screening study in newborn children and preschool children (n = 342). The analysis included a parallel estimation of alpha$_1$-AT activity by three procedures: on gelatin gel (Billingsley and Cox, 1980) by radial immunodiffusion (Mancini *et al.*, 1965) and by double-rockets immunodiffusion (Laurell, 1972). Diminished alpha$_1$-AT activity probably equivalent to the heterozygous phenotypes, was discovered in about 5% of the children examined. The screening was continued (Kotulska-Hofmokl *et al.*, 1982), and until 1982 more than one thousand newborn children were examined. The percentage of alpha$_1$-AT deficients remained nearly the same. In seven of the newborn children, the deficiency reached very low levels (about 20%), thus indicating a possible Pi^Z homozygotic phenotype. On the basis of these results, the frequencies of alpha$_1$-AT deficiency in the Polish population and of particular Pi phenotypes appear quite similar to frequencies seen in other European countries.

11.4.2 Changes in alpha$_1$-AT due to external agents

Several environmental agents may cause a decrease in alpha$_1$-AT activity. The inhibitory action of cadmium has been well documented in vivo and in vitro (Marek *et al.*, 1980). The effects of lead and organic solvents in occupational exposures have been demonstrated (Cwynar-Bialobrzeska, 1987). In addition, our own results suggest that smoking causes an increase of alpha$_1$-AT in serum (Cwynar-Bialobrzeska, 1987).

Susceptibility to disease was confirmed epidemiologically in people with deficient serum alpha$_1$-AT levels (or defective phenotypes). After two years of observations, four deaths were seen in children with alpha$_1$-

AT deficiency (Kotulska-Hofmokl *et al.*, 1982). A high percentage (20%) of surviving children with alpha$_1$-AT deficiency had acute, chronic or current respiratory tract diseases. These findings have been also confirmed by unpublished data on Polish children. In a group of subjects with chronic kidney disease (n = 21), Cwynar-Bialobrzeska (1987) observed significantly lower concentrations of alpha$_1$-AT in serum. The antitrypsin activity of serum was also decreased as compared to that of the control group. In 1986, Rujner *et al.* described three cases of neonatal hepatitis connected with Pi^Z alpha$_1$-AT phenotypes. Parents of these children represented heterozygous phenotypes Pi^{MZ}, and siblings were characterized by phenotypes Pi^Z, Pi^{MZ} or Pi^M, which confirms the autosomal recessive heredity of the trait.

In a large study comprising more than 700 cases, the high percentage (18%) of low serum alpha$_1$-AT levels (200 mg%) was observed in a group of Polish children with chronic liver diseases (Children Health Centre in Warsaw, unpublished data).

11.4.3 Methodological problems

(a) Two-phase screening

Between 1978 and 1980, we carried out several projects to evaluate two screening models for detecting the presence of pathological alpha$_1$-AT phenotypes:

1. pre-employment examinations related to occupational exposures to chemicals causing emphysema in people with pathological alpha$_1$-AT phenotypes;
2. more widespread screening for deficiency of alpha$_1$-AT among newborn children.

Both models used a two-phase screening procedure. During the first phase the examination measured alpha$_1$-AT levels in serum; the second phase was devoted to phenotying.

For economic reasons, the first phase was divided into two stages. The analysis for the first stage was inexpensive. During the second stage, a more sophisticated immunological procedure was used. To choose appropriate tests, the correlations between particular methods were examined and the results published (Pilacik, 1977; Pilacik and Kowalczyk, 1979; Cwynar and Pilacik, 1981; Pilacik, 1987). We evaluated the assay of Billingsley and Cox (1980), the screening test of Laurell (1972), the Mancine *et al.* immunological test (1965) and the colorimetric method of Eriksson (1965).

For Pi phenotyping, electrophoretic separation on starch gel (Fagerhol, 1968) and the antigen-antibody crossed electrophoresis of Laurell (1965) were compared. The latter could be recommended for screening $alpha_1$-AT deficits in both adult workers and newborn children.

(b) Relation between $alpha_1$-AT concentration and activity

Using the accessible human material, we tried to answer the question of whether the increase in $alpha_1$-AT concentration is equivalent to the rise of antitrypsin activity of the serum. The problem is important for the proper choice of screening methods; some of them measure $alpha_1$-AT concentration, others measure the $alpha_1$-AT activity. The data published by Cwynar-Bialobrzeska (1987) indicate that a good correlation between the two types of procedure can be obtained only in screening of newborn children, using cord blood. However, in adults, perhaps as the result of environmental agents, we frequently observed some disproportion between the $alpha_1$-AT concentration and the $alpha_1$-AT activity in serum. The results obtained in smokers can serve as a good example. This situation favours screening of $alpha_1$-AT deficiencies performed immediately after birth.

11.5 METHAEMOGLOBIN REDUCTASE

Some rare variants exist of the $NADH^+$-dependent enzyme that protects haemoglobin from oxidation. Low activities seen in some variants result in methaemoglobinemia. The pharmaco- and ecogenetic role of the enzyme is not known.

11.5.1 Frequency of methaemoglobin reductase deficiency in the Polish population

In 1968, we tried to establish the approximate frequency of red blood cell methaemoglobin reductase deficiency in the Polish population (Hanke and Linkowska-Pinkiewicz, 1968). Among 100 individuals of the general population, no clear deficiency could be found by the procedure described by Scott and Griffith (1959). In 1975, the trial was repeated on a population sample consisting of 349 men and 51 women not exposed occupationally to noxious agents. The analysis of the distribution curve of methaemoglobin reductase exhibited a highly significant right asymmetry. No significant difference in the dispersion between men and women could be observed. One could expect an activity smaller than 50% of geometrical value with a probability higher than 1% (Hanke *et al.*, 1975).

11.5.2 Changes in methaemoglobin reductase activity due to external agents

The activity of methaemoglobin reductase dependent on reduced NAD is much lower in newborn children than in adults. This difference is the reason for methaemoglobinemia in children after ingestion of some drugs. In psoriatic patients, one could expect a decrease of activity of red blood cell enzymes participating in the generation of a reduced form of NAD. Measurement of methaemoglobin reductase in red blood cells of 20 psoriatic patients (10 women and 10 men) between 15 and 72 years of age, (Kaczorowska-Hanke *et al.*, 1973) showed that the earlier observed increase in G–6-PD activity (dependent on the reduced form of NADP) (Kaczorowska-Hanke *et al.*, 1971) represents a compensation for the reduced haemoglobin defence mechanism. This is normally based chiefly on methaemoglobin reductase, which in psoriatic patients seems to be lower than normal. However, this decrease was not large enough to cause methaemoglobinemia after a normal dose of phenacetine (3 × 150 mg daily) for three successive days.

11.6 *N*-ACETYLTRANSFERASE

Liver *N*-acetyltransferase is important for the acetylation of many drugs (e.g. izoniazid). Individuals can be divided into two groups, 'rapid inactivators' and 'slow inactivators', depending on the effectiveness of the acetylation process and, hence, of inactivation of the drug (Evans and Clarke, 1980). No significant differences have been observed between slow and rapid inactivators; however, some side effects have been noticed. In family studies the difference between slow and rapid inactivators is genetically determined. Recently, a higher tendency to develop urinary bladder cancer as an effect of industrial exposure to some aromatic amines has been documented for slow inactivators (Cartwright *et al.*, 1982; Hanssen *et al.*, 1985).

11.6.1 Frequency of rapid and slow inactivators in the Polish population.

In 1981, Skretkiewicz *et al.* published a study on the frequency of rapid and slow inactivators in the central part of Poland. Ninety people (41 men and 49 women) between 18 and 65 years of age were examined using the method of Eidus *et al.* (1973). The hydrazide of isonicotinic acid was used as a marker drug. Among the inhabitants of the city of Lodz, 37.8% were found to be rapid inactivators, 44.4% slow inactivators and 17.8% intermediate inactivators. When the intermediate inactivator group is combined with the slow inactivators, these results are very similar to those obtained by Pawlowska and Rakowska (1976) in a group of 300 tuberculosis patients (71 women and 229 men); in this study, sulfamidine was a marker drug and the procedure of Bratton and Marschall (1939) was applied.

11.6.2 *N*-acetyltransferase and cancer

We started a retrospective study and a follow-up study to evaluate the hypothesis that slow inactivators are more susceptible to bladder cancer when exposed to aromatic amines. The frequencies of slow inactivators were compared in two groups. One consisted of patients with diagnosed bladder cancer but without occupational contact with aromatic amines. The other group consisted of bladder cancer patients occupationally exposed to aromatic amines (such as benzidine). The results of this study are presented in Table 11.1

Table 11.1 Assessment of inactivation phenotype in male patients (age 66.2 ± 9.2 years) suffering from bladder cancer

		Slow	*Intermediate*	*Rapid*
I	Patients with bladder cancer not exposed to aromatic amines (n = 43)[a]	26 (60.5%)	8 (18.6%)	9 (20.9%)
II	Patients with bladder cancer exposed to aromatic amines (n = 24)	21 (87.6%)	1 (4.1%)	2 (8.3%)
III	Control group (n = 22)	10 (45.5%)	3 (13.7%)	9 (40.9%)

[a]Examinations performed in cooperation with the Department of Pharmacology of Medical Academy of Lodz.
Statistical evaluation:
I/II P = 0.018 (significant)
I/III P = 0.107 (nonsignificant)
II/III P = 0.0028 (highly significant).

The difference between the frequency of slow inactivators in the two groups is statistically significant (P = 0.018). If intermediate inactivators are included with the slow inactivators, the percentage of the whole group is higher than the one seen in the general population of the central part of Poland.

So far, the results of retrospective studies unanimously support the thesis about the higher frequency of slow inactivators among the patients suffering from urinary bladder cancer who were previously exposed to aromatic amines. However, we are not fully convinced by these data that people who are slow inactivators are at higher risk of bladder cancer when exposed to aromatic amines. There are two primary objections: first, the disease process, especially the therapeutic agents, could perhaps influence the acetylation rate in the liver; and second, the long-term exposure to aromatic amines may not be harmless to the liver and may influence the acetylation rate.

These reservations can only be examined in a prospective study. For this purpose, a study was initiated of all workers exposed to benzidine until 1981 in a dye manufacturing trade. The benzidine production stopped in 1981. The working conditions during the preceding period are described in detail elsewhere (Krajewska *et al.*, 1980). The total group, including people currently exposed to benzidine in a dye factory in Wola Krzysztoiporska, comprised 90 workers with an average age of 47 years and an exposure period of about 22 years. There were 60 (66.7%) slow inactivators, 10 (11.1%) intermediate inactivators and 20 (22%) rapid inactivators. The percentage of slow inactivators appears to be higher than among the general population in Lodz. One may not conclude from these data that long-term exposure to aromatic amines does not influence the inactivation phenotype. To resolve this problem, we are starting another study to assess the inactivation rate of workers exposed to aromatic amines, including both new and long-term employees. In this new study we shall try also to resolve the problem of the influence of cancer therapy on the acetylation rate of the liver. Other reports (for example Bulovskaya *et al.*, 1978) show that the acetylation rate during cancer therapy can be changed in both directions, that is from slow to rapid and from rapid to slow. Tumour growth itself may also influence the acetylation rate.

11.7 CONCLUSIONS

1. Human populations are exceedingly diverse. These variations are caused partly by differences in the environmental conditions in which they live, but they also depend on inborn differences.
2. For preventative purposes, it is extremely important to be able to define the most sensitive members in a population, all the more so as hygienic standards do not protect all workers who exhibit susceptibility higher than normal.
3. However, social and ethical aspects constitute serious arguments against genetic screening.
4. Impetus must be given to scientific research and epidemiological studies to clarify the genetic background of hypersusceptibility and

to gather information on the prevalence of important polymorphisms in different populations.

5. A major research challenge is the development of accurate, quick and inexpensive tests to identify heterozygous carriers of traits.
6. The aim of laboratory quality assurance is to reduce laboratory errors as much as possible and to ensure compatibility of analytical results from different laboratories.
7. Every result of a mass screening should be exploited only for benefit and not against the person under examination.

REFERENCES

Aldrich, J. E., Fernandez, J. R. and Tunis, J. J. (1974) A screening method for measuring G-6-PD activity, in *Enzymology in the Practice of Laboratory Medicine* (eds Ph. Blume and E. Freier), Academic Press, New York and London, pp. 271–90.

Berger, L. (1963) The semi-quantitative determination of the activity in red cells or other fluid, Tentative Technical Bulletin No. 400, Sigma Chemical Company, St Louis.

Beutler, E. (1957) The glutathione instability for drug-sensitive red blood cells. A new method of in vitro detection of drug sensitivity. *J. Lab. Clin. Med.*, **49**, 84–95.

Beutler, E., and Balude, M. (1963) Methemoglobin reduction. Studies of the interaction between cell population and the role of methylene blue. *Blood*, **22**, 323–33.

Beutler, E., Dern, A. and Alving, A. (1955) The hemolytic effect on erythrocytes of primaquine. VI. An *in vitro* test for the sensitivity of erythocytes to primaquine. *J. Lab. Clin. Med.*, **45**, 40–5.

Billingsley, G. D. and Cox, D. W. (1980) Functional assay of alpha$_1$-antitrypsin in obstructive lung diseases. *Am. Rev. Respir. Dis.*, **121**, 161–2 (modified).

Bratton, A. C. and Marschall, E. K. (1939) A new coupling component for sulphanilamid determination. *J. Biol. Chem.*, **123**, 537–50.

Brewer, G. J., Tarlov, A. R., Alving, A. S. (1960) Methemoglobin reduction test. *Bull. WHO*, **22**, 633–40.

Bulovskaya, L. N., Krupkin, R. G., Bochina, T. A. *et al.* (1978) Acetylator phenotype in patients with breast cancer. *Oncology*, **35**, 185–8.

Carson, P. E., Flanagan, C. I., Ickes, C. E. and Alving, A. S. (1956) Enzymatic deficiency in primaquine-sensitive erythrocytes. *Science*, **124**, 484–95.

Cartwright, R. A., Rogers, H. J., Barham-Hall D. *et al.* (1982) Role of *N*-acetyltransferase phenotypes in bladder carcinogenesis: a pharmacogenetic epidemiological approach to bladder cancer. *Lancet*, **ii**, 842–7.

Cwynar, B. and Pilacik, B. (1981) Trial for improving the analytical procedure for detection of alpha$_1$-antitrypsin deficiency in subjects entering the work in a Polish industrial plant. *Pol. Tyg. Lek.*, **31**, 1169–71 (in Polish).

Cwynar-Bialobrzeska, B. (1987) The methodological fundamentals for alpha$_1$-antitrypsin screening in epidemiological research, Medical Academy, Lodz (dissertation in Polish).

Dabrowski, Z., Miszta, H. and Marszalek, K. (1980) Characteristic of glucose-6-phosphate dehydrogenase after intoxication with heavy metals and acetylphenylhydrazine. *Folia Med. Cracov.*, **3–4**, 411–18 (in Polish).

Eidus, L., Varughese, P., Hodgkin, M. M. *et al.* (1973) Simplification of isoniazid phenotyping procedure to promote its application in the chemotherapy of tuberculosis. *Bull. World Health Org.* **49**, 507–16.

Eriksson, S. (1965) Studies in alpha$_1$-antitrypsin deficiency. *Acta Med. Scand.*, **177**, (Suppl. 432), 421–8.

Evans, D. A. P. and Clarke C. A. (1980) Pharmacogenetics. *Br. Med. Bull.*, **17**, 234–8.

Fagerhol M. K. (1968) The Pi-system. Genetic variants of serum alpha$_1$-antitrypsin. *Ser. Haematol.*, **1**, 153–60.

Fairbanks, V. F. and Beutler, R. (1962) Simple method for detection of erythrocyte glucose–6-phosphate dehydrogenase deficiency (G–6-PD Spot Test). *Blood*, **20**, 291–5.

Flieger, S., Karon, H., Sarnowska, I. and Warchol, T. A. (1980) Trial using determination of glucose–6-phosphate dehydrogenase and reduced glutathione in erythrocytes as indicators of oral and maxillary neoplasm. *Czasopismo Stomatologiczne*, **12**, 1087–91 (in Polish).

Hanke, J. (1968) Methemoglobin reductase, glucose–6-phosphate dyhydrogenase and glutathione peroxidase activity in workers chronically exposed to the action of methaemoglobin forming compounds. *Proceedings from the National Congress of Occupational Medicine*, Warsaw (unpublished).

Hanke, J. (1984) Genetic susceptibility to toxic substances and its relationship to cancerogenesis, *Monitoring Human Exposure to Carcinogenetic and Mutagenetic Agents*, (eds A. Berlin, M. Draper, K. Hemminki and H. Vainio), IARC Sci. Publ. No. 59, International Agency for Research on Cancer, Lyon.

Hanke J., Krajewska B. and Andryszek Cz. (1975) Distribution of the red blood cell enzymes activity in the population not exposed professionally to the action of chemical agents. *Medycyna Pracy*, 3, 255–259 (in Polish).

Hanke, J., Kuczynski, T., Linkowska, R. (1968), An attempt of evaluation of frequency of G–6-PD deficiency in red cells among Polish population, *Med. Pr.*, **6**, 532–6 (in Polish).

Hanke, J. and Linkowska-Pinkiewicz, R. (1968) Methaemoglobin reductase ($NADH_2$:lipoamide oxidoreductase) its significance and the method of its determination. *Diagnostyka Laboratoryjna*, **4**, 321–6 (in Polish).

Hanssen, H. P., Agarwal, P. P., Goedde, H. W. *et al.* (1985) Association of *N*-acetyltransferase polymorphism and environmental factor with bladder carcinogenesis. Study of north German population. *Eur. Urol.*, **11**, 263–6.

Kaczorowska-Hanke, H., Ruszczak, Z. and Hanke, J. (1973) The activity of $NADH_2$:lipoamide oxidoreductase in red blood cells in patients with psoriasis. *Przeglad Dermatologiczny*, **1**, 11–15 (in Polish).

Kaczorowska-Hanke, H., Ruszczak, Z., Linkowska, R. and Hanke, J. (1971) Activity of red blood cells glucose–6-phosphate dehydrogenase in psoriatics. *Przeglad Dermatologiczny*, **3**, 267–72 (in Polish).

Krajewska D., Adamiak-Zieba, J., Suwalska D. and Kostecka K. (1980) Evaluation of occupational benzidine exposure of those operating automatized benzidine production. *Med. Pr.*, **31**, 403–10 (in Polish).

Kleinhauer, E. and Betke, K. (1963) Elution procedure of the demonstration of methemoglobin in red cell of human blood smears. *Nature*, **199**, 1196–7.

Kotulska-Hofmokl, J., Chojnowska-Miszewska, W., Pilacik, B. and Nowak S. (1982) Alpha$_1$-antitrypsin deficiency in neonatal population and follow up of children for 1–2 years. *Pediatr. Pol.*, **10**, 765–70 (in Polish).

Laurell, C. B. (1965) Antigen-antibody crossed electrophoresis. *Anal. Biochem.*, **10**, 358.

Laurell, C. B. (1972) A screening test for alpha$_1$-antitrypsin deficiency. *Scand. J. Clin. Lab. Invest.*, **29**, 247–8.

Laurel, C. B. and Eriksson, S. (1963) The electrophoretic alpha$_1$-globulin pattern of serum in alpha$_1$-antitrypsin deficiency. *Scand. J. Clin. Lab. Invest.*, **15**, 132–8.

Linkowska-Pinkiewicz, R. (1979) Deficient activity of glucose–6-phosphate dyhydrogenase in erythrocytes of foreign students starting studies in Poland. *Pol. Tyg. Lek.*, **4**, 137–9 (in Polish).

Linkowska-Pinkiewicz, R. (1980) Selection of methods and analysis of their usefulness in detection of glucose–6-phosphate dehydrogenase deficits in erythrocytes. *Diagnostyka Laboratoryjna*, **5**, 325–30, (in Polish).

Löhr, G. W. and Waller, H. D. (1970) Glucose–6-phosphate dehydrogenase, in *Methods of enzymatic Analysis* (ed. H. D. Bermeyer), Academic Verlag, Berlin, pp. 599–606.

Mancini, G., Carbornare, C. O. and Heremans, J. F. (1965) Immunochemical quantitations of antigens by single radial immunodiffusion. *Immunochemistry*, **2**, 235–40.

Marek, K., Wocka-Marek, T. and Marks, J. (1980) Effects on occupational exposure to cadmium on the activity of alpha$_1$-antitrypsin in the serum. *Pol. Arch. Med. Wewn.*, **63**, 53–6 (in Polish).

Martin, J. P., Vandeville, D., Martin, C. and Ropartz, C. (1974) Identification of phenotypes of the Pi system and comparison of methods of tracing deficits in alpha$_1$-antitrypsin. *Ann. Biol. Chem.*, **32**, 197–207.

Opolska, B. (1974) Studies on the occurrence of Pi system phenotypes in the population of southern Poland. *Przegl. Lek.*, **31**, 851–4, (in Polish).

Pawlowska, I. and Rakowska, Z. (1976) Determination of the phenotype of INH inactivation in patients with pulmonary tuberculosis using an indirect method and a model sulphamethazine. *Pneumonol. Pol.*, **XLVI**, (8) 605–11 (in Polish).

Pilacik, B. (1977) Screening methods for the detection of alpha$_1$-antitrypsin deficiency in blood serum. *Med. Pr.*, **2**, 143–7 (in Polish).

Pilacik, B. (1987) Alpha$_1$-antitrypsin phenotypes and their role in medical diagnosis. *Med. Pr.*, **5**, 376–83 (in Polish).

Pilacik, B. and Kotulska-Hofmokl, Z. (1979) A trial of application of gelatin test for screening of newborn and children for detection of alpha$_1$-antitrypsin deficiency. *Pediatr. Pol.*, **1**, 17–20 (in Polish).

Pilacik, B. and Kowalczyk, J. (1979) Laboratory procedures for the detection of deficiencies and determination of alpha$_1$-antitrypsin phenotypes. *Med. Pr.*, **3**, 207–12 (in Polish).

Rujner, J., Pilacik, B., Rosikiewicz, E., and Wozneiwicz, B. (1986) Alpha$_1$-antitrypsin deficiency as a cause of neonatal hepatitis. *Pediatr. Pol.*, **12**, 808–13 (in Polish).

Sass, M. D., Caruso, C. J. and Axelrod, D. R. (1966) Rapid screening for D-glucose–6-phosphate: NADP oxidoreductase deficiency with methylene blue. *J. Lab. Clin. Med.*, **68**, 156–62.

Scott, E. M. and Griffith, I. V. (1959) Methaemoglobin reductase. *Biochim. Biophys. Acta*, **34**, 584–7.

Skretkiewicz, J., Mazurowa, A. and Orszulak, D. (1981) Acetylation phenotype in a group of population from the Lodz region. *Pol. Tyg. Lek.*, **3**, 89–91 (in Polish).

Triger, D. R., Millwandsedler, G. M., Czaykowska, A. A. *et al.* (1976) Alpha$_1$-antitrypsin deficiency and liver disease in adults. *Q. J. Med.*, **45**, 351–72.

Tysper, Z. and Karon, J. (1971) Determination of glucose–6-phosphate dehydrogenase in red cells. *Diagnostyka Laboratoryjna*, **7**, 47–51 (in Polish).

Wierzbowska, A. and Hanke, J. (1968) A trial of using 'G–6-P D spot Test' screening investigations in the Polish population. *Pol. Tyg. Lek.*, **3**, 95–6 (in Polish).

12 Phenotypes of human serum esterases reacting with organophosphates, carbamates and other esters

V. Simeon and E. Reiner

Institute for Medical Research and Occupational Health, University of Zagreb, Zagreb, Croatia Yugoslavia

12.1 SUMMARY

Organophosphates and carbamates react in human serum with only a few enzymes, either inhibiting them or being destroyed by them. The enzymes involved are cholinesterase (EC 3.1.1.8) and A(ryl)-esterases (A-esterase, paraoxonase, EC 3.1.8.1 and arylesterase, EC 3.1.1.2). Interindividual differences of biochemical properties of these enzymes have been established in many populations and are explained by the existence of different genotypes.

Serum cholinesterase occurs as the usual form, the atypical and several other variants. Ninety-five per cent of the population belongs to the usual genotype, while all other variants occur with frequencies of less than 5%. Cholinesterase variants differ in their susceptibilities to positively charged organophosphates and carbamates; these are stronger inhibitors of the usual cholinesterase than of the atypical enzyme. Paraoxonase, the most extensively studied A(ryl)-esterase, occurs in three genotypes which differ in their hydrolytic activities; about 50% of the individuals belong to the low-activity genotype. These individuals may be at a higher risk if exposed to organophosphorous (OP) compounds. To predict the genetic predisposition of an individual to the adverse effects of these compounds, the sensitivity and overall detoxification capacity should be examined together with the genotype.

12.2 INTRODUCTION

Three enzymes in human serum react with carbamates and carboxylic acid esters: cholinesterase (EC 3.1.1.8), aryldiakylphosphatase (paraoxonase, A-esterase) (EC 3.1.8.1) and arylesterase (EC 3.1.1.2). For the latter two enzymes, a common name A(ryl)-esterases is used throughout this chapter. These enzymes are present in a number of phenotypes that have been demonstrated in most of the populations studied. Individuals having different phenotypes of the enzymes respond differently to the above compounds.

The mechanism of the reaction between cholinesterase and OP, carbamates and carboxylic acid esters is well understood; the three-step mechanism includes a Michaelis-type complex and the acylated enzyme. With this mechanism, the reactions of substrates and inhibitors are analogous, differing only in the rates of deacylation.

The mechanism of the reaction between A(ryl)-esterases and OP, carbamates and carboxylic acid esters is still unknown. All three groups of compounds are considered to be substrates of A(ryl)-esterases. However, the substrate specificity of the EC 3.1.8.1 and EC 3.1.1.2 enzymes is still not clear. The enzyme-substrate Michaelis complex shows up in the kinetics of substrate hydrolysis, but no other intermediates have been demonstrated. Choline esters and other esters, bearing a positive charge, do not usually react with A(ryl)-esterases.

The physiological role of both serum cholinesterase and A(ryl)-esterases is unknown. Some conditions, inherited or acquired, can produce increasing or decreasing cholinesterase activities (Whittaker, 1986). Low cholinesterase activity may be due to inherited abnormalities or to such conditions as liver disease and pregnancy. Increased cholinesterase activity may be due to obesity and some diseases or, very rarely, may be genetically determined. Low A(ryl)-esterase activities have been observed following myocardial infarction, in disorders of lipid metabolism and in patients with cystic fibrosis.

12.3 CHOLINESTERASE

Many reports have discussed the distribution of serum cholinesterase activities in different populations. Some data obtained using three common substrates are listed in Table 12.1. The distribution of activities has been found to be unimodal and symmetrical, with interindividual coefficients of variation between 17 and 28%, regardless of the substrate and of differing experimental conditions, such as temperature and pH of assay. The ratios between the highest and lowest activities range between 2.2 and 8.5. Generally, women below 40 years of age fall into the lower range of activities.

Individuals with variant phenotypes of serum cholinesterase are not readily detected in the distribution curve of a randomly chosen population, although they are usually associated with a lower than normal activity. This observation is due to the fact that they occur with a frequency of less than 5%, while 95% belong to the usual phenotype. Data obtained on a large Caucasian population (Evans, 1986) and a population group from Zagreb (Simeon *et al.*, 1987; Simeon and Prester, 1987) are shown in Table 12.2.

Two genetic loci (E_1 and E_2) determine the cholinesterase activity in plasma. Locus E_1 controls most of the genetic variants. The best known are usual (U), atypical (A), fluoride-resistant (F) and silent (S) cholinesterase; all combinations of heterozygous variants occur. On the basis of inhibition with dibucain and sodium fluoride, ten cholinesterase genotypes can be recognized; in addition, K and J variants have been found (Evans, 1986).

The most pronounced differences in biochemical properties are seen between the usual and atypical variants, the cholinesterase phenotypes for which the amino-acid sequences are known (Lockridge and La

Table 12.1 Cholinesterase activities in human sera measured with acetylthiocholine (ATCh), butyrylthiocholine (BTCh) or propionylthiocholine (PTCh).

Substrate (mM)	*Activity range ($\mu mol\ min^{-1} ml^{-1}$)*	*CV*[a]	*n*	*References*
ATCh (1)	0.35–2.80	23	252 (M)	Wilhelm and
	0.35–2.45	25	148 (F)	Bradamante (1980)
ATCh (1)	0.89–3.62	23	134 (M)	This chapter
	0.89–3.24	26	103 (F)	
ATCh (5)	1.8–4.4	18	36 (M)	Huizenga *et al.*
	1.6–3.5	17	41 (F)	(1985)
ATCh (5)	0.82–5.23	24	180 (M)	Simeon *et al.* (1987)
	1.29–4.14	25	154 (F)	
BTCh (6)	2.27–7.40	28	226 (M)	Prellwitz *et al.*
	2.05–6.70	28	198 (F)	(1976)
BTCh (7)	1.74–14.82	—	715 (M)	Blaauwen *et al.*
	2.20–9.38	—	147 (F)	(1983)
PTCh (1)	1.63–6.27	24	134 (M)	This chapter
	13.0–5.46	25	103 (F)	
PTCh (2)	3.10–7.50	21	221 (M)	Prellwitz *et al.*
	2.90–6.90	21	173 (F)	(1976)

[a]CV is the coefficient of interindividual variation for *n* individuals (M = male, F = female).

Table 12.2 Frequency of some cholinesterase phenotypes and the relative mean enzyme activities (RMA) for the substrates: acetylthiocholine (ATCh), benzoylcholine (BzCh) and propionylthiocholine (PTCh). (A) Data from Evans (1986) were obtained on a large Caucasian population and (B) data from Simeon *et al.* (1987) and Simeon and Prester (1987) are on a population group from Zagreb

Phenotype[a]	*A*		*B*		
	Frequency (%)	*RMA PTCh*	*Frequency %*	*RMA ATCh*	*RMA BzCh*
U	95	100	91	100	100
A	0.04	23	—	—	—
Some heterozygotes					
US	0.5	51	—	—	—
UA	3.3	70	4.5	69	64
UF	0.5	80	3.3	93	90
AK	0.4	43	0.8	54	51

[a]Phenotype abbreviations: U: usual, A: atypical, S: silent, F: fluoride resistant, K: variant named after W. Kalow.

Du, 1978; McGuire *et al.*, 1989). The two enzyme molecules differ in only one amino acid out of a total of 574. The usual phenotype has an aspartic acid in position 70 while the atypical has glycine in that position. This difference appears to account for the altered affinity towards positively-charged compounds. The atypical phenotype has a very much lower affinity for positively-charged compounds, while the affinity for uncharged compounds is about the same for both phenotypes. The susceptibilities of atypical and usual cholinesterases towards some organophosphates and carbamates are presented in Table 12.3. The ratio of inhibitor concentration producing 50% enzyme inhibition (A/U) is 1.0 for uncharged compounds and higher for charged compounds. The highest ratio was found for the charged compound Ro 02–0683.

Table 12.3 Inhibition of usual (U) and atypical (A) cholinesterase (Kalow and Davis, 1958). The I_{50} is the inhibitor concentration producting 50% enzyme inhibition

Inhibitor	$I_{50}(A)/I_{50}(U)$
Organophosphates	
DFP	1.0
TEPP	1.0
Paraoxon[a]	1.0
Phosphostigmine[a]	47
VX[a]	21
Carbamates	
Neostigmine	25
Physostigmine	18
Ro 02–0683	120

[a]V. Simeon and M. Škrinjarić-Špoljar, unpublished data.

The muscle relaxant succinylcholine (suxamethonium) has a positive charge and is very slowly hydrolyzed by the atypical cholinesterase at pharmacological concentration. The in vivo action of this drug is therefore prolonged, and paralysis of up to several hours' duration may occur in individuals homozygous for the atypical cholinesterase variant (Evans, 1986; Whittaker, 1986). In addition, individuals with activities at the lower end of the activity distribution curve are also at increased risk, irrespective of their phenotype and of whether this lower activity is inherited or acquired. Abnormal cholinesterase variants are also resistant to inhibition by the local anaesthetic agent dibucain.

12.4 A(RYL)-ESTERASES

The existence of several A(ryl)-esterase phenotypes has long been known, although fewer studies have been conducted with these enzymes than with cholinesterases. Two substrates are most often used in studies of the A(ryl)-esterases: paraoxon and phenylacetate. All population studies have shown that the distribution of paraoxonase activities is either bimodal or trimodal (Geldmacher-von Mallincrodt and Diepgen, 1988; Reiner *et al.*, 1989a,c). In most populations, about 50% of individuals belong to the group having a low paraoxonase activity, while the remainder are distributed between the medium- and high-activity phenotypes. The same distribution of activities has also been shown for other organophosphates such as Armin, Methyl-paraoxon, Chlorthion-oxon and EPN-oxon (Geldmacher-von Mallincrodt and Diepgen, 1988). Data on A(ryl)-esterase activities measured in our laboratory using different substrates on a random population from Zagreb are shown in Table 12.4. Distribution studies using carbamates as substrates have not been carried out as far as we know. Some data on the hydrolysis of carbamates obtained in our laboratory are shown in Table 12.4.

Many studies have been conducted on the hydrolysis of phenylacetate. All have shown that the distribution of activities is unimodal and symmetrical (Geldmacher-von Mallincrodt and Diepgen, 1988; Reiner et al 1989a,c). However, using beta-naphthylacetate, we have shown that the distribution in a random population sample from Zagreb is unimodal but not symmetrical (Reiner *et al.*, 1989b).

The hydrolysis of OP compounds by A(ryl)-esterases is considered to contribute to the detoxification of these highly potent compounds, which are used as pesticides and also as poisons in chemical warfare. It has been suggested that individuals with high A(ryl)-esterase activities will be more resistant to the effects of OP compounds because hydrolysis of OP will decrease the phosphorylation rate of cholinesterases. In vitro studies with paraoxon have shown that cholinesterase is less inhibited

Table 12.4 A(ryl)-esterase activities in human serum

Substrate (mM)	*Activity range ($\mu mol\ min^{-1} ml^{-1}$)*	*n*	*Distribution*	*References*
Organophosphates				
Paraoxon (5)	0.0069–0.75	382	Polymodal	Reiner *et al.* 1989b,c
Armin (6)	0.091–0.286	46	Polymodal	Škrinjarić-Špoljar and Reiner, 1968
N-methyl carbamates				
Sevin (0.8)	0.03–0.09	12	—	Reiner and Škrinjarić-Špoljar, 1968
Propoxur (0.8)	0.01–0.03	12	—	
Carboxylic acid esters				
Phenylacetate (5)	33–404	382	Unimodal	Reiner *et al.*, 1989b,c
Beta-naphthylacetate (1)	5.9–45.4	145	Unimodal	Reiner *et al.*, 1989b

in serum samples that have a high paraoxonase activity (Eckerson and La Du, 1984). To enable prediction of the capacity of the whole body to detoxify OP compounds, more data are needed on enzymes from other tissues that also participate in these reactions.

12.5 CONCLUSION

In some countries, individuals with atypical cholinesterase, sensitive to the muscle relaxant succinylcholine, are registered and patients are adequately instructed on the risks associated with succinylcholine application. An atypical cholinesterase is less inhibited by positively-charged organophosphates or carbamates than a usual enzyme. In spite of that, individuals with atypical cholinesterase or other variants may be unusually sensitive to OP compounds (Prody *et al.*, 1989). In addition, due to the lower affinity of the atypical enzyme for positively-charged compounds, it can be expected that such individuals are more refractant to therapy by pyridinium oximes. Individual differences in the capacity of the serum to hydrolyse OP compounds have been established, but the genetic predisposition concerning differences in the overall detoxification capacity remains to be evaluated. Mandatory monitoring of exposure of workers handling OP compounds and recognition of early symptoms of exposure may identify individuals more sensitive to the compounds than the average person (USEPA, 1988).

REFERENCES

Blaauwen, H. D. von den, Poppe, W. A. and Tritschler, W. (1983) Cholinesterase (EC 3.1.1.8) mit Butyrylthiocholin-iodid als Substrate: Referenzwerte in Abhängigkeit von Alter und Geschlecht unter Berücksichtigung hormonaler Einflüsse und Schwangerschaft (Cholinesterase (EC 3.1.1.8) with butyrylthiocholine iodide as substrate: age- and sex-dependent reference values with special reference to hormonal effects and pregnancy). *J. Clin. Chem. Clin. Biochem.*, **21**, 381–6.

Eckerson, H. W. and La Du, B. N. (1984) A mathematical model for evaluating the reaction of paraoxon with human serum cholinesterase and with polymorphic forms of paraoxonase. *Drug Metab. Dispos.*, **12**, 57–62.

Evans, R. T. (1986) Cholinesterase phenotyping: clinical aspects and laboratory applications. *CRC Crit. Rev. Clin. Lab. Sci.*, **23**, 35–64.

Geldmacher-von Mallinckrodt, M. and Diepgen, T. L. (1988) The human serum paraoxonase – polymorphysm and specificity, *Toxicol. Environ. Chem.*, **18**, 79–196.

Huizenga, J. R., Belt, K. van der and Gips, C. H. (1985) The effect of storage at different temperatures on cholinesterase activity in human serum. *J. Clin. Chem. Clin. Biochem.*, **23**, 283–5.

Kalow, W. and Davis, R. O. (1958) The activity of various esterase inhibitors towards atypical human serum cholinesterase. *Biochem. Pharmacol.*, **1**, 183–192.

Lockridge, O. and La Du, B. N. (1978) Comparison of atypical and usual human serum cholinesterase. *J. Biol. Chem.*, **253**, 361–6.

McGuire, M. C., Nogueira, C. P., Bartles, C. F. *et al.* (1989) Identification of structural mutation responsible for the dibucaine-resistant (atypical) variant from human serum cholinesterase. *Proc. Natl Acad. Sci. USA*, **86**, 953–7.

Prellwitz, W. von, Kapp, S. and Müller, D. (1976) Vergleich von Methoden zur Activitätbestimmung der Serumcholinestrasen (Acylcholine-acylhydrolase E.C. 3.1.1.8) und deren diagnostische Wertigkeit (Comparative methods for the determination of the activity of serumcholinesterases (acylcholine-acylhydrolase EC 3.1.1.8) and their diagnostical value). *J. Clin. Chem. Clin. Biochem.*, **14**, 93–7.

Prody, C. A., Dreyfus, P., Zamir, R. *et al.* (1989) De novo amplification within a 'silent' human cholinesterase gene in a family subjected to prolonged exposure to organophosphorus insecticides. *Proc. Natl Acad. Sci. USA*, **86**, 690–4.

Reiner, E., Aldridge, W. N. and Hoskin, F. C. G. (eds) (1989a) *Enzymes hydrolysing organophosphorus compounds*, Ellis Horwood, Chichester.

Reiner, E., Radić, Z., Pavković, E. and Simeon, V. (1989b) Substrate specificity of human serum A-esterases. International meeting on carboxylesterases, Osnabrück. *Biol. Chem. Hoppe-Seyler*, **370**, 987–8.

Reiner, E., Radić, Z. and Simeon, V. (1989c) Hydrolysis of paraoxon and phenylacetate by human serum esterases, in *Enzymes Hydrolysing Organophosphorus Compounds* (eds E. Reiner, W. N. Aldridge, and F. C. G. Hoskin), Ellis Horwood, Chichester, pp. 30–40.

Reiner, E. and Škrinjarić-Špoljar, M. (1968) Hydrolysis of some monomethylcarbamates in human sera. *Croat. Chem. Acta*, **40**, 87–90.

Simeon, V., Buntić, A., Šurina, B. and Flegar-Meštrić, Z. (1987) Cholinesterase phenotyping and distribution of activity in sera of 346 individuals. *Acta Pharm. Jugosl.*, **37**, 107–14.

Simeon, V. and Prester, Lj. (1987) Kinetics of inhibition of human serum cholinesterase phenotypes with dimethylcarbamate Ro 02–2083. Seventh Meeting of the Yugoslav Association of Clinical Biochemists, Novi Sad, *Jugoslav. Med. Biokem.*, **6**, 214.

Škrinjarić-Špoljar, M. and Reiner, E. (1968) Hydrolysis of diethyl-p-nitrophenylphosphate and ethyl-p-nitrophenyl-ethylphosphonate by human sera. *Biochim. Biophys. Acta*, **165**, 289–92.

USEPA (1988) USA Environmental Protection Agency, Worker Protection Standards for Agricultural Pesticides: Public Meetings and Proposed Rule, 40 CFR Parts 156 and 170, 25996–26008, *Federal Register, Part XII.*

Whittaker, M. (1986) Cholinesterase, in *Monographs in Human Genetics* (ed. L. Beckman), Vol. 11, Karger, Basel.

Wilhelm, K. and Bradamante, V. (1980) Blood cholinesterase activity in workers exposed to anticholinesterases, a ten-year follow-up. *Arh. Hig. Rada Toksikol.*, **31**, 109–24.

13 Predisposition in exposure to carbon disulfide

D. Djuric

Institute of Occupational and Radiological Health, Belgrade, Yugoslavia

13.1 SUMMARY

In occupational exposure to carbon disulfide (CS_2), some workers are hypersusceptible to the toxic effects of this chemical. This hypersusceptibility is associated with a decreased metabolism of Antabuse, another sulfur-containing compound. In CS_2-poisoned patients, the decreased metabolism persists for several years after cessation of exposure, possibly due to irreversible damage to the mixed-function oxidases in the liver. Thus, a decreased capacity for detoxification of CS_2 leads to toxicity that includes further decrease in the detoxification capacity. The hypersusceptibility is likely to be inherited, but the pattern of inheritance is unknown.

13.2 INTRODUCTION

Healthy human subjects exhibit large variations in response to toxic substances and other xenobiotics. The magnitude of these toxicokinetic variations in subjects of the same age and sex, under similar environmental conditions, ranges from 4-fold to 40-fold; more than 20 different host factors (such as genetic constitution, age, sex, health status, diet, way of life) are responsible for these variations, which interact dynamically (Vesell, 1982).

A dose-response curve for a toxicant in a human or animal population will usually indicate a susceptible group of the subjects on one side and a resistant group on the other side. The distribution curve may take the form of a symmetrical or asymmetrical, sometimes Gaussian, curve.

13.3 HYPERSUSCEPTIBILITY TO CS_2

We were able to study this situation in a viscose factory, where workers were exposed to CS_2. In 1963, a team of experts from the Institute of Occupational and Radiological Health in Belgrade started an intensive, broadly-based study on biochemical, toxicological and medical aspects of CS_2 toxicology in exposed workers. During these studies the problem of individual susceptibility to this toxicant emerged very early on.

First, a reliable exposure test for workers exposed to CS_2 had to be identified. Determination of CS_2 in blood and urine did not produce reliable data on the level of exposure due to the existence of the free and bound fractions. In 1963, Vasak, *et al.* published a paper on a new exposure test for CS_2: the iodine-azide test (IAT), based on a finding of Yoshida (1955) that the reaction between sodium azide and iodine is catalyzed by CS_2 metabolites present in the urine of exposed animals:

$$2NaN_3 + I_2 \rightarrow 3N_2 + 2NaI$$

Vasak *et al.* (1963) measured decoloration time (t) in seconds that elapsed from the moment of adding the reagent to the urine. They also determined the urine creatinine concentration (c) to calculate the exposure coefficient E:

$$E = c \times \log t$$

In 1964, we used the IAT to establish the level of exposure of workers in our viscose factory and obtained a very good correlation with the average CS_2 concentration in the air (Djuric *et al.*, 1965). We observed that one group of workers (the recovery group) showed a normal level of E before work and an increased level at the end of exposure each day. A second group of workers (the non-recovery group) showed an increased level of E at the beinning of work and a highly increased level at the end of the exposure. Subsequently, Stokinger and Mountain (1967) suggested the use of the non-recovery pattern of IAT for detection of hypersusceptible workers.

Continuing our study, we established that the non-recovery pattern represents an early sign of biochemical disorder and that this group of workers will develop symptoms of chronic poisoning on further exposure (Graovac-Leposavic *et al.*, 1967; Djuric *et al.*, 1968). We suggested that the non-recovery pattern is a consequence of a biochemical defect in the enzyme systems metabolizing sulfur compounds, caused by CS_2. Accordingly, the non-recovery pattern represents an early sign of poisoning.

To evaluate this hypothesis, it was necessary to find a suitable sulfur-

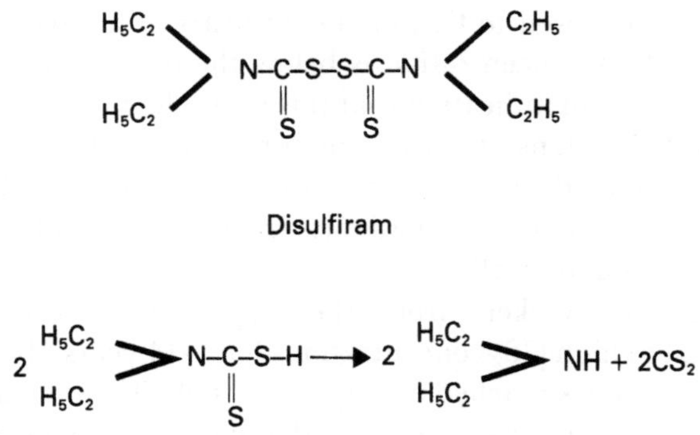

Figure 13.1 Disulfiram and metabolites

containing compound that can be used to determine the rate of metabolism of sulfur compounds in the humans. For this purpose we selected the tetraethylthiuram disulfide (TETD, Antabuse, disulfiram), which is used for the treatment of alcoholism. Antabuse is metabolized to produce two molecules of diethyldithiocarbamate (DDC) and finally two molecules of diethylamine and two molecules of CS_2 (Fig. 13.1).

Skalicka (1967) had previously determined the excretion of DDC in the urine of alcoholics treated with Antabuse. We combined the results of DDC and IAT in urine of 22 chronic alcoholics treated with Antabuse (Novak *et al.*, 1968). After oral administration of 0.5 g Antabuse, we sampled 6-hour urine fractions. The DDC excretion showed great variation from 5.2 to 771 μg/g creatinine (mean value 158.4 DDC μg/g creatinine).

The differences in DDC excretion of chronic alcoholics treated with Antabuse could be ascribed partly to differences in absorption, but mainly to individual differences in the rate of metabolism of this compound; this factor is probably genetically influenced. We speculated that Antabuse could be used for evaluation of the individual rate of metabolism of sulfur compounds and estimation of individual susceptibility to CS_2.

13.4 DEVELOPMENT OF THE ANTABUSE TEST

We decided to administer Antabuse orally to various groups of viscose workers (Djuric *et al.*, 1973). For this purpose we selected three groups.

1. Eighteen workers from the viscose yarn production. These workers

had been exposed to CS_2 for 11–15 years at an average level of 32 mg/m^3. This concentration is below the maximum permitted level (60 mg/m^3), an indication that this group had not been dangerously exposed. No signs of chronic poisoning were observed. This group was considered as a 'control' group or 'normal DDC excretion' group, similar to the patients treated for alcoholism and not exposed to CS_2 at all.

2. Twenty-one workers from the staple-cell operation who were exposed to high CS_2 concentrations for 7–14 years. The CS_2 concentration at times reached average values of 300–400 mg/m^3. Selected workers of this group were under constant medical surveillance and did not show any signs of poisoning. Therefore, they were considered to represent the 'resistant' group of workers.
3. Thirty-three workers who worked in the staple-cell operation at one time under the same conditions as group 2. During the exposure they developed signs of chronic poisoning (polyneuritis) on one or two occasions and were removed from exposure. They had worked since removal for 5–10 years without CS_2 exposure (for example as gardeners or guards). They were partially disabled and constituted the 'susceptible' group.

Diethyldithiocarbamate was determined in a 6-hour urine sample. The results of DDC determination in the urine of the three groups of workers defined above are presented in Table 13.1. The statistical evaluation of results and the correlation between groups are also presented. The mean level of DDC excretion in the resistant group (90.0 μg/g creatinine) is lower than in the control group (160.5 μg/g creatinine), but this difference is not statistically significant due to the wide variation in the values obtained in the controls. Neither group showed any signs of CS_2 poisoning: the resistant group, in spite of increased exposure, the control group, because of low exposure. The mean excretion of DDC is lowest in the susceptible group (49.7 μg/g creatinine); this level is statistically highly significant when compared both to the controls and the resistant group.

These results suggest that high exposure to CS_2 causes a disturbance in the metabolism of sulfur compounds. This implies that CS_2 damages its own enzymic system, which also metabolizes other sulfur compounds. Susceptible and resistant workers worked in the same operation exposed to high CS_2 concentrations (300–400 mg/m^3). In the susceptible workers, CS_2 may have injured the enzyme system metabolizing sulfur compounds to less toxic metabolites. This injury appears to be irreversible. In the resistant group, CS_2 caused a decrease in the metabolism of sulfur compounds, but not below a certain threshold at which metabolic conversion

Table 13.1 Determination of diethyldithiocarbamate (DDC) in urine after oral administration of 0.5 g disulfiram

Group	*No. of workers*	*Duration of exposure (y)*	*DDC (μg/g creatinine) Range*	*Mean* ± SD
Control	18	11–15	50–564	160.5 ± 130.7
Resistant	21	7–14	35–130	90.0 ± 39.3[b]
Susceptible	33	[a]	30–110	49.7 ± 29.0[c,d]

[a]Removed from CS_2 exposure for 5–10 years
[b]$p<0.05$ when compared to control group (+−test)
[c]$p<0.001$ when compared to control group(+−test)
[d]$p<0.001$ when compared to control group(+−test)

to the less toxic metabolities is prevented. On the basis of these findings, we suggested the use of the Antabuse test as a pre-employment screening for a worker's suitability for the viscose factory. Individuals with a high metabolic capacity for Antabuse (high DDC excretion) would be expected to represent the resistant group, and individuals with a low DDC excretion to represent the susceptible group. Cooper (1973) included this test in his review on indicators of susceptibility to industrial chemicals.

In the meantime we isolated and identified in the urine of viscose workers all three CS_2 metabolites responsible for the iodine-azide reaction: mercaptothiazolinone (Pergal *et al.*, 1972), thiocarbamide (Pergal *et al.*, 1972) and thiocarbamidemercaptothiazoline carbamic acid (Pergal *et al.*, 1977a). We also developed a method for the quantitative determination of thiocarbamide (thiourea) in the urine of exposed workers (Pergal *et al.*, 1977b).

13.5 STUDIES BY OTHER TEAMS

In Bulgaria, Pavlova and Usheva (1976) monitored the DDC excretion in 35 healthy persons after oral administration of Antabuse. They sampled 4 × 6-hour urine samples (24 hours) after oral intake of Antabuse. The first and second urine samples showed almost identical DDC concentration and a significant decrease was seen in the following two urine samples. The IAT was increased in parallel with the increase in DDC excretion.

In another Yugoslav viscose factory, Besarabic (1978) administered Antabuse to 61 workers who had been exposed to CS_2 for 4 to 5 years. She compared the level of DDC excretion with data on total and specific absenteeism. From the results obtained, she concluded that a high rate of absence due to CS_2-related morbidity appeared in workers with a DDC excretion below 150 μg/g creatinine. Workers with a DDC excretion above this value appeared less susceptible to the effects of CS_2. Besarabic suggested that workers should only be accepted for work involving CS_2 exposure if their DDC excretion was above 150 μg/g creatinine during the administration of a pre-employment Antabuse test. Based on the accumulated evidence, the Antabuse test was cited in the WHO publication on CS_2 (WHO, 1979).

In Poland, Podobinski (1983) administered Antabuse to 96 workers in a viscose factory. He confirmed our observations that DDC excretion decreases with the level of exposure but not with the duration of exposure. Podobinski suggested that an adaptation to CS_2 exists due to the induction of an enzymic system metabolizing sulfur compounds. Podobinski also compared DDC excretion with the simultaneous estimation of the liver damage using the Bengal rouge I^{131} test. The results suggested that long-term exposure to low concentrations of CS_2 may cause toxic effects mainly in the liver, thus damaging its metabolic function. Podobinski recommended the Antabuse test for the detection of changes in the metabolism of CS_2 that probably precede the development of symptoms of chronic poisoning.

13.6 DISCUSSION

An increased exposure to CS_2 could well injure the enzyme systems that metabolize sulfur compounds (including CS_2 and disulfiram). Excretion of metabolites would then be retarded or decreased. This suppression of sulfur compound metabolism would represent an early sign of CS_2 toxicity or even an early sign of poisoning. If the primary toxic substance (i.e. CS_2) and/or intermediates are more toxic than the final metabolites, suppression of their detoxification would result in intoxication. In exposed workers the great variation observed in the capacity of this metabolic system is linked to susceptibility or resistance to the toxic agent. Thus, in the resistant group the metabolism of sulfur compounds under CS_2 influence is not decreased below a certain threshold. Therefore this group shows no signs of poisoning throughout the years of high exposure. On the other hand, under the same exposure a majority of susceptible workers reached the metabolic threshold after shorter or longer exposure, and developed symptoms of poisoning.

In the disabled workers, this very low metabolic conversion persists even 5 to 10 years after the end of exposure to CS_2. One potential explanation for this observation is that the metabolic disturbance caused by CS_2 is irreversible in nature. We also observed that some of these workers, when intoxicated for the first time, soon showed early signs of intoxication when they returned to the same exposure after recovery. Thus they were then moved to non-exposed workplaces.

To explain this phenomenon we have to consider the metabolism of CS_2 and its effects on the enzyme system metabolizing sulfur compounds. After absorption into the human organism, 10–30% of absorbed CS_2 is exhaled and partly eliminated via the skin. The retained 70–90% undergoes biotransformation. The metabolites produced, together with less than 1% of unchanged CS_2, are excreted in the urine or exhaled (CO_2) (WHO, 1979; Beauchamp *et al.*, 1983). The metabolism of CS_2 is not fully elucidated, and a tentative scheme is presented in Fig. 13.2. The two main pathways are:

1. binding to (a) glutathione (GSH) to produce thiazolidine–2-thione–4-carboxylic acid and 2-oxothiazolidine–4-carboxylic acid, and (b) amino acids to produce dithiocarbamates and 2-thio–5-thiazolidine;
2. oxidative desulfuration performed by microsomal mixed-function oxidases (MFO).

The quantity of metabolites produced by the first pathway (including thiourea) is relatively small. Therefore, from the quantitative as well as the toxicological point of view, the second pathway is the most important for CS_2 conversion.

Microsomal MFO are the most important enzyme systems involved in biotransformation of xenobiotics in the liver and other organs of mammalian and other species (Parke, 1981; Djuric 1988). These processes include oxidation, hydroxylation, desulfuration and dehalogenation. The oxidative desulfuration of CS_2 is probably mediated by MFO in two stages. In the first stage, a sulfur atom is liberated and carbonyl sulfide, as an intermediate, is formed. In the second stage, the second sulfur atom is liberated to produce CO_2 as a final product. Unstable intermediates also produce thiourea under the influence of cytosolic carbonic anhydrase. Due to the lack of substrate specificity of MFO, several other sulfur-containing compounds, such as disulfiram (Antabuse), diethyldithiocarbamates, thiocarbonyl and thiourea, are also undergoing oxidative desulfuration. Reactive sulfur atoms liberated during this process may become bound to the cellular components and initiate toxic changes in the liver and other tissues.

The apoprotein of cytochrome P–450 (which is the central enzyme molecule of MFO) may be the primary target of the toxic action of liberated sulfur atoms (WHO, 1979; Schenkman and Kupfer, 1982; Djuric, 1988). A portion of the sulfur atoms released during the oxidative desulfuration processes may react with the -SH groups of cysteine residues in the cytochrome P–450 apoprotein. The protein moiety of cytochrome P–450 would then represent the main target of the toxic action

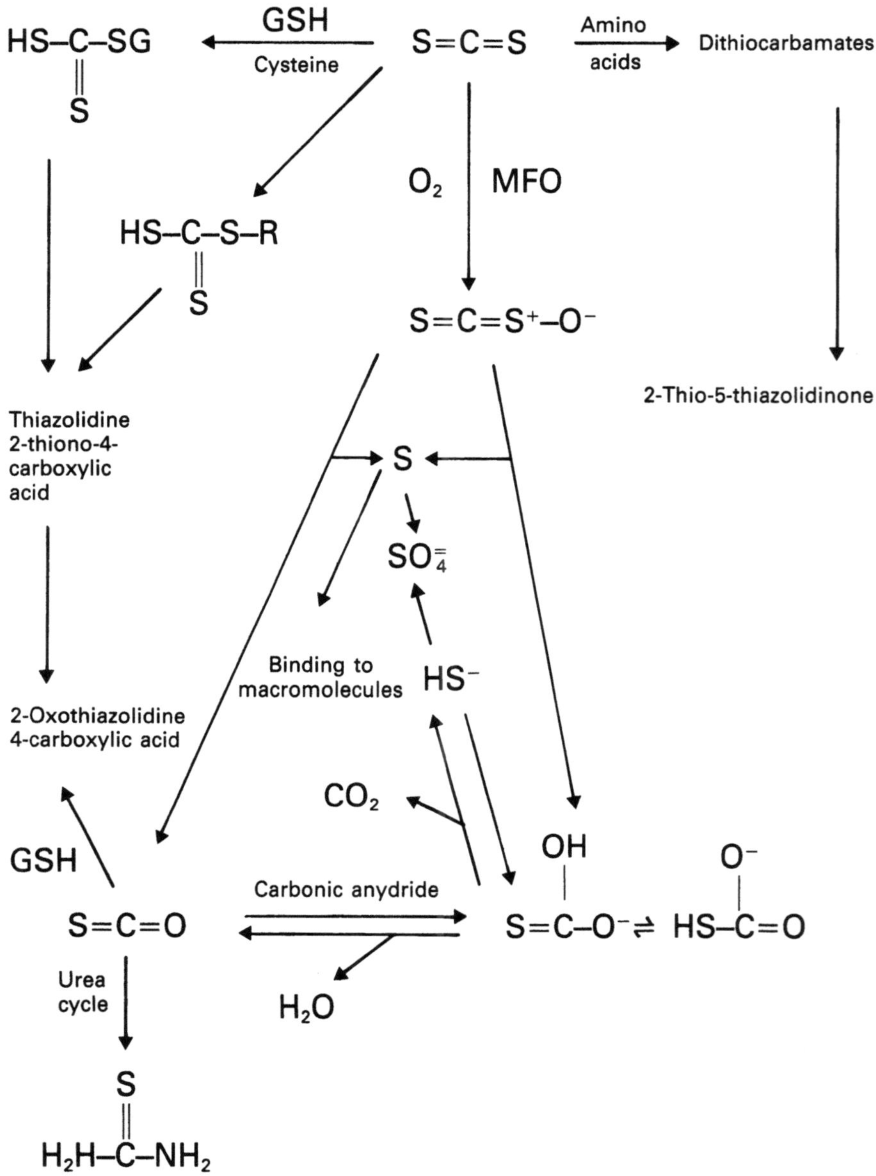

Figure 13.2 Proposed pattern of CS_2 metabolism (after Beauchamp *et al.*, 1983)

of CS_2 (WHO, 1979; Parke, 1981; Schenkman and Kupfer, 1982; Beauchamp *et al.*, 1983; Djuric, 1988).

Small concentrations of CS_2 are likely to produce an inducing effect on MFO, as seen with numerous other toxicants. However, higher CS_2 concentrations will exert an inhibitory effect on MFO as described. A small increase in the dose can therefore produce an effect much larger than expected, with steep elevation in tissue concentrations in subjects with a reduced capacity for metabolic detoxification.

Presumably, the subjects with greater metabolic capacity of MFO represent the resistant group. When this subgroup is exposed to higher concentrations of CS_2, they are relatively quickly able to detoxify the absorbed dose in spite of partial inactivation of the cytochrome P–450 enzyme. Subjects with a lower metabolic capacity of MFO will represent the hypersusceptible subgroup, in whom the destruction of enzyme molecules will cause a serious decrease in the rate of metabolism, so that intoxication will ensue (Djuric, 1979, 1984). The persistent low metabolic capacity of MFO in some subjects many years after cessation of CS_2 exposure can be explained by an irreversible disturbance of MFO. The exact nature of this phenomenon is unknown.

Interindividual toxicokinetic variations are mainly based on differences in MFO and other metabolic activity, rather than on differences in absorption, distribution or excretion of the toxicant (Breimer and Danhof 1980; Vesell, 1982, 1985). The synthesis, capacity and function of enzymes are inherited characteristics. Therefore, genetic factors would appear to constitute the main cause of large interindividual variations in toxicokinetics.

The differences in capacity of MFO for oxidative desulfuration of sulfur-containing compounds probably represent an inheritable variation. As a consequence, the Antabuse test could represent a predictive test of hypersusceptibility in workers exposed to CS_2 and other sulfur-containing compounds. By applying this test we can at least establish the metabolic capacity of the MFO responsible for metabolism of sulfur-containing compounds. Therefore, the Antabuse test represents an attempt to identify biochemical phenotypes associated with hypersusceptibility to effects of sulfur-containing chemicals.

The next step would be to develop a pedigree analysis of individuals and their families to reveal the role of genetic factors in controlling interindividual variations in the MFO system responsible for metabolizing sulfur-containing compounds. Family studies could be used to establish the Mendelian mode of transmission of this condition.

REFERENCES

Beauchamp, Jr R. O., Bus, J. S., Popp, J. A. and Boreiko, C. J. (1983) A critical review of the literature on carbon disulfide toxicity. *CRC Crit. Rev. Toxicol.*, **11**, 162–278.

Besarabic, M. (1978) 'Antabuse test' and absenteeism in workers exposed to CS_2 *Arh. Hig. Rada Toksikol.*, **29**, 323–6.

Breimer, D. D. and Danhof, M. (1980) Interindividual differences in pharmacokinetic and drug metabolism, in *Towards Better Safety of Drugs and Pharmaceutical Products* (ed. D. D. Breimer), Elsevier/North-Holland Biomedical Press, Amsterdam.

Cooper, C. (1973) Indicators of susceptibility to industrial chemicals. *J. Occup. Med.*, **15**, 355–9.

De Matteis, F. (1982) Loss of microsomal components in drug-induced liver damage, in cholestasis and after administration of chemicals which stimulate heme catabolism, in *Hepatic Cytochrome P–450 Monooxygenase System*, (eds J. B. Schenkman and D. Kupfer), Pergamon Press, London, pp. 307–40.

Djuric, D. (1979) Antabuse test as an indicator of susceptibility to CS_2. *G. Ital. Med. Lav.*, **1**, 49–52.

Djuric, D. (1984), Proposals for the 'Antabuse test', *G. Ital. Med. Lav.*, **1**, 49–52.

Djuric, D. (1988) Molecular-cellular aspects of occupational exposure to toxic chemicals. Part I, Toxicokinetics, World Health Organization, Geneva (unpublished document OCH/87.2).

Djuric, D., Graovac-Leposavic, Lj. and Rezman, I. (1968) The use of IAT in occupational hygiene and medicine. *Arh. Hig. Rada Toksikol.*, **19**, 408–12 (in Yugoslav).

Djuric, D., Postin-Grujin, A., Graovac-Leposavic Lj. and Delic V. (1973) Disulfiram as an indicator of human susceptibility to CS_2. *Arch. Environ. Health*, **15**, 287–92.

Djuric, D., Surducki, N. and Berkes, I. (1965) Iodine-azide test on urine of persons exposed to CS**2**. *Br. J. Ind. Med.*, **22**, 321–3.

Graovac-Leposavic, Lj., Djuric, D., Pavlovic, A. and Jovicic, M. (1967) The use of iodine-azide test for early diagnosis of CS_2 poisoning in *Toxicology of Carbon Disulfide* (eds H. Brieger and J. Teisinger), Excerpta Medica Foundation, Amsterdam, pp. 52–61.

Novak, Lj., Djuric, D. and Fridman, V. (1968) Urinary excretion of sulphur compounds in persons treated with disulfiram. *Arch. Environ. Health*, **19**, 408–12.

Parke, D. V. (1981) The endoplasmic reticulum, in *Drug Metabolism*, Part B (eds P. Jenner and B. Treska), Dekker, New York, pp. 1–52.

Pavlova, S. and Usheva, G. (1976) Excretion of DDC in healthy persons due to intake of disulfiram, Proceedings of Bulgarian Congress on Occupational Health, Sofia, pp. 62–4 (in Bulgarian, unpublished).

Pergal, M., Kobilarov, N., Raletic, J. *et al.* (1977b) Determination of small quantities of thicarbamide in urine. Egypt. *J. Occup. Med.*, **5**, 251–7.

Pergal, M., Vukojevic, N., Cirin-Popov, N. *et al.* (1972) Carbon disulfide metabolites excreted in the urine of exposed workers, I. Isolation and identification of 2-mercapto–5-thiazolinone. *Arch. Environ. Health*, **25**, 38–41.

Pergal, M., Vukojevic, N. and Djuric, D. (1972) Isolation and identification of thiocarbamide. *Arch. Environ. Health*, **25**, 42–4.

Pergal, M., Vukojevic, N. and Djuric, D. (1977a) Identification of the third metabolite giving positive iodine-azide reaction. Abstracts of the IV International Symposium on Occupational Health in the Production of Artificial Fibres, Helsinki, Finland, 6–10 June 1977, pp. 29–30 (unpublished).

Podobinski, M., (1983) Evaluation attempt of quasi-genetic factors affecting xenobiotic Antabus metabolism in CS_2 industrial exposure, *Health Effects of Combined Exposure to Chemicals in Work and Community Environment*, Interim Document 11, WHO Regional Office for Europe, Copenhagen, pp. 399–414.

Schenkman, J. B. and Kupfer, D. (eds) (1982) *Hepatic Cytochrome P-450 monoxygenase system*, Pergamon Press, Oxford.

Skalicka, B. (1967) Sulphur compounds in urine after treatment with TETD. *Prac. Lek.*, **19**, 408–12 (in Czechoslovakian).

Stokinger, H. E. and Mountain, J. T. (1967) Progress in detecting the workers hypersusceptible to industrial chemicals. *J. Occup. Med.*, **9**, 537–42.

Vasak, V., Vanececk, M. and Kimmelova, B., (1963) The assessment of exposure of workers to CS_2 by the iodine-azide test. *Prac. Lek.*, **15**, 145–53.

Vesell, E. S. (1982) On the significance of host factors that affect drug disposition. *Clin. Pharmacol. Ther.*, **31**, 1–7.

Vesell, E. S. (1985) Implications for risk assessment of host factory causing large pharmacokinetic variations. *Toxicol. Ind. Health*, **1**, 133–52.

WHO (1979) *Carbon Disulfide*, Environmental Health Criteria No. 10, World Health Organization, Geneva.

Yoshida, K. (1955) Fate of CS_2 in the body-formation of thioketone bonds. *J. Sci. Labour (Tokyo)*, **31**, 209–213 (in Japanese).

14 Genetic susceptibility to DNA-methylating agents

H. W. Rüdiger

Unit of Toxicogenetics, Institute of Occupational Health, University of Hamburg, Hamburg, Germany

14.1 SUMMARY

The individual risk of developing cancer from genotoxic agents is determined by both the exposure and the ability to cope with the genotoxic burden. Among the cellular defence mechanisms, DNA repair is particularly important. This is demonstrated by the increased cancer incidence in several genetic disorders that are characterized by a specific defect in DNA repair. However, the individual variation in DNA repair capacity could conceivably affect the risk of common cancers as well, which may arise as a consequence of specific genotoxic influences. This was in fact demonstrated recently by an inverse relation between the incidence of human lung cancer and the activity of O^6-methylguanine-DNA-methyltransferase. This ubiquitous cellular protein removes small alkyl groups from the O^6-position of guanine, the latter being considered to be a most critical site in DNA with respect to the occurence of permanent DNA modification.

14.2 INTRODUCTION

DNA is a reactive molecule; it is biologically designed to be so. DNA has to be copied, rearranged and transcribed, and all these processes are triggered in a living cell by molecules that chemically react with specific sites of the DNA. Conceivably, this ever-present burden of DNA modifications can only be tolerated because of powerful mechanisms that permanently undo unwanted alterations of the genetic material. It can further be argued that in the presence of an enormous amount of endogenously occurring DNA modifications the DNA repair capacity, rather than the additional mutagenic load from the environment, determines the incidence of mutations. The test cases are the genetic diseases ataxia telangiectasia (Painter and Young, 1980), Bloom's syndrome (Willis and Lindahl, 1987) and xeroderma pigmentosum (Kraemer *et al.*, 1975; Cleaver, 1983), each of which is characterized by a defect in a specific DNA repair function. Patients with such an autosomal recessively inherited condition are a 100-fold or more at risk of developing cancer as compared to the general population.

A lower but still enhanced cancer incidence was also demonstrated in otherwise asymptomatic heterozygote carriers of these Mendelian disorders (Swift, 1976). Genetic predisposition to cancer, however, is not confirmed to a few rare genetic conditions but has been observed for almost any common type of cancer in humans (Table 14.1), and variations of individual DNA repair capacities are expected to contribute significantly as well (Lindahl, 1987). Although the exact number of individual repair mechanisms responsible for the integrity of the human cellular genome is not known (Lindahl, 1982) and only a few can be tested as yet, some experimental proof already indicates that the concept is valid. Repair of O^6-methylguanine-DNA is one example.

Table 14.1 Genetic cancer risk: familial incidence of common tumours

Site of cancer	*Relative risk of first degree relatives of a patient*
Breast	1.3–9
Colon	5
Malignant melanoma	8.5
Lung	2.8–5.2
Uterus	> 16% familial cases
Stomach	3
Prostata	3

14.3 BIOCHEMICAL AND BIOLOGICAL PROPERTIES OF O^6-METHYLGUANINE-DNA-METHYLTRANSFERASE

DNA may be methylated by various endogenous and exogenous genotoxic substances. Among these, nitrosamines have been most extensively studied (Lijinsky, 1986). Several are carcinogenic in a variety of animal species. The various DNA-methylating agents may exhibit different site-specific methylating properties (Day *et al.*, 1987) and may thereby produce very different frequencies of mutation events.

This has been experimentally demonstrated (Doniger *et al.*, 1985) with the three different methylating agents *N*-methyl-N^1-nitro-*N*-nitrosoguanidine (MNNG), *N*-methyl-*N*-nitrosourea (MNU) and methyl methansulfonate (MMS). Although each induced transformation of Syrian hamster embryo cells in a dose-dependent manner, MNNG was 100-fold more effective than MNU and 500-fold more than MMS. However, concentrations that produced equal amounts of O^6-methylguanine (O^6-mgua) cell survival and transformation frequencies were similar (Table 14.2).

Table 14.2 Quantitative comparison of biological and biochemical responses to DNA-methylating agents (from Doniger *et al.*, 1985)

				*μmol/mol guanine**	
Methylating agent	*Concentration* (μM)	Survival %	% Transf'd colonies	N^7-mgua*	O^6-mgua*
MNNG	1.7	88	1.4	57	4.5
MNU	100	91	1.0	62	5.0
MMS	1000	85	1.5	1900	5.0

*N^7-mgua: N^7-methylguanine; O^6-mgua: O^6-methylguanine

Thus methylation of the O^6-position of guanine is particularly critical with respect to the generation of permanent DNA alterations (Loveless, 1969; Pegg, 1984). Consequently, this site is protected by a distinct repair mechanism, the O^6-methylguanine-DNA-methyltransferase (O^6-mgua-MT) which exclusively removes methyl groups from the O^6-position of guanine (Yarosh *et al.*, 1984, 1985). Numerous studies have demonstrated the presence of O^6-mgua-MT in almost all tissues (Pegg, 1978; Waldstein *et al.*, 1982; Scudiero *et al.*, 1984; Domoradzki *et al.*, 1985; Brent, 1986; Maher *et al.*, 1986; Morimoto *et al.*, 1986). This repair protein has some interesting properties.

- The repair process is entirely error free, because it removes the methyl group without an interruption of the integrity of the DNA strand.
- The cleaved methyl group is transferred to the methyltransferase protein which is thereby irreversibly inactivated. This implies that the amount of guanine-O^6 methylations that can be repaired equals the amount of transferase molecules in a cell. Consequently, even small variations of transferase activity will cause parallel variations of repair capacity in a cell.
- The primary event responsible for induction of surplus O^6-mgua-MT de novo synthesis is the transitory inactivation of existing transferase molecules by covalently protein-bound methyl groups.

14.4 IN VITRO STUDIES IN HUMANS

Various experimental studies with human cells in culture have convincingly demonstrated that cellular sensitivity to DNA-methylating chemicals is dependent on the presence and the activity of O^6-mgua-MT.

- Individual mutagenicity and genotoxicity are related to the constitutive MT activity (Domoradzki *et al.*, 1984; Scudiero *et al.*, 1984).
- Depletion of MT activity increases the cellular sensitivity to mutations (Domoradzki *et al.*, 1985), to cytotoxicity (Babich and Day, 1987) and to SCE induction (Trey and Gerson, 1987).
- The transferred human MT gene restores methylation resistance in mutant cell strains that lack O^6-mgua-MT (Bignami *et al.*, 1987).
- The adaptive response induces MT activity and resistance to DNA methylation (Frosina and Abbondandolo, 1985).

O^6-mgua-MT exhibits considerable interindividual variability in humans. This was tested in human cultured fibroblasts from 110 donors. Briefly, the technique was as follows (Rüdiger *et al.*, 1989). The cultures were initiated by standard techniques, homogenized by brief sonication in a N_2-atmosphere and incubated thereafter with a known amount of O^6-mgua containing DNA oligonucleotides. After acid hydrolysis of the sample, O^6-mgua was separated by HPLC with an acetonitrile/ammonium formate gradient of 7–10.5% at nucleosil SA 10 and detected fluorimetrically at an excitation wavelength of 286 nm and an emission wavelength of >360 nm. The reduction of O^6-mgua as compared to controls without cell extract represents the cellular repair capacity. The result is presented in Fig. 14.1. Surprisingly, there is a greater than 20-fold interindividual difference in O^6-mgua-MT in the 110 fibroblast cultures tested. Thus, O^6-mgua-MT represents a biologically important DNA repair mechanism which shows a considerable interindividual difference. It seems logical to assume that these differences in repair capacities would represent individually different sensitivities to DNA-methylating agents, and that this sensitivity will be related to a person's cancer risk.

14.4.1 Studies in cancer patients

In order to test this hypothesis three groups of probands were studied: 45 individuals with histologically-confirmed lung cancer, 39 individuals with malignant melanoma and 29 normal healthy controls without a personal or a familial history of cancer. The cancer patients were tested prior to onset of a specific therapy. Lung cancer patients were used because of the established correlation between the occurrence of this tumour and cigarette smoke (Wynder and Graham, 1985). Cigarette

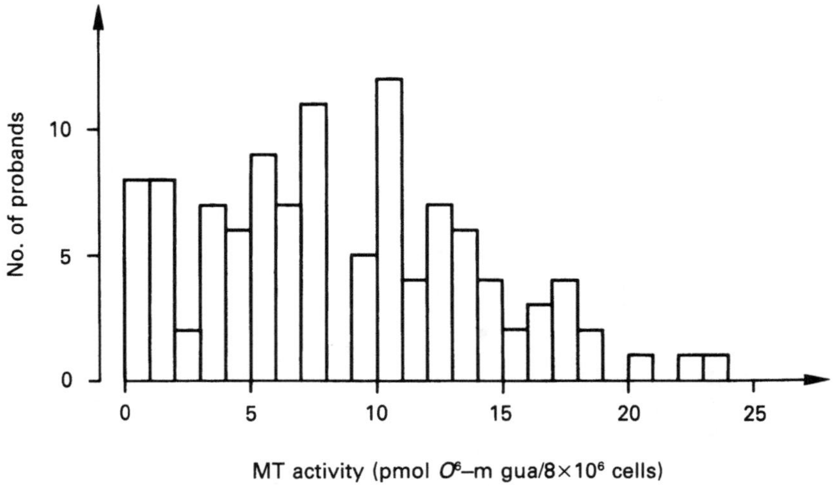

Figure 14.1 Distribution of O^6 methylguanine-DNA-methyltransferase (O^6-mgua-MT) activity in fibroblast cultures of 110 probands.

smoke contains large amounts of tobacco-specific nitrosamines that have been demonstrated to cause cancer in laboratory animals (Belinsky *et al.*, 1987; Hecht and Hoffmann, 1988). Patients with a cutaneous malignant melanoma were used as a control group since this type of tumour is not likely to have a connection with exposure to DNA-methylating chemicals, but rather to an exposure to UV light (Kripke, 1979; Elwood *et al.*, 1985). The cellular enzymic activity of methyltransferase was not affected by UV light. The distribution of methyltransferase activities in the three groups of probands tested is shown in Fig. 14.2. Lung cancer patients exhibit a significantly lower repair of O^6-mgua as compared to melanoma patients and normal healthy controls. Of particular interest, four probands had no detectable transferase activity at all and all these belonged to the lung cancer group.

14.4.2. In vitro studies with cells from MT-proficient and -deficient probands

The reduced O^6-mgua repair of lung cancer patients is consistent with the hypothesis that the basal activity of this repair system may be related to an individual's sensitivity to DNA-methylating agents. If this hypothesis is correct, fibroblast cultures with different repair capacities should exhibit accordingly a different sensitivity to the genotoxic action of nitrosamines.

In order to confirm this, the four fibroblast cultures with no detectable

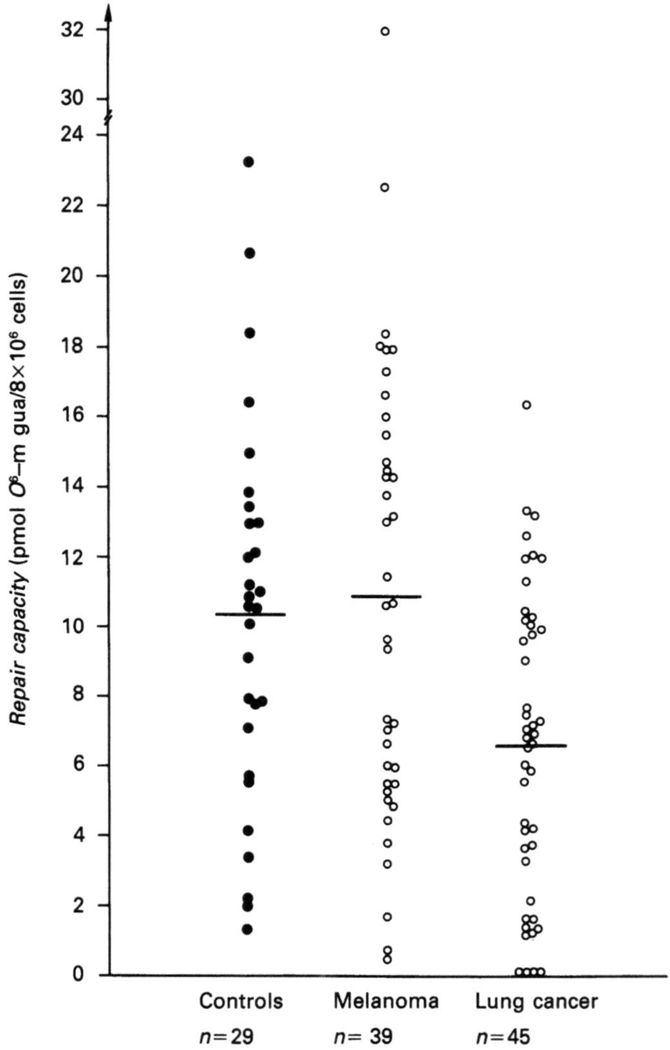

Figure 14.2 Activity of O^6 methylguanine-DNA-methyltransferase (O^6-mgua-MT) in fibroblasts from 29 normal healthy controls, 39 patients with cutaneous malignant melanoma and 45 patients with lung cancer.

repair capacity and four cultures with normal activity of O^6-mgua-MT were exposed to varying amounts of MNNG up to 10 μmol/l, and the induction of sister chromatid exchange and micronuclei was determined. The result demonstrates that fibroblasts with low O^6-mgua-MT are more sensitive than those with a normal repair capacity (Fig. 14.3).

Analogous experiments were performed with the tobacco-specific

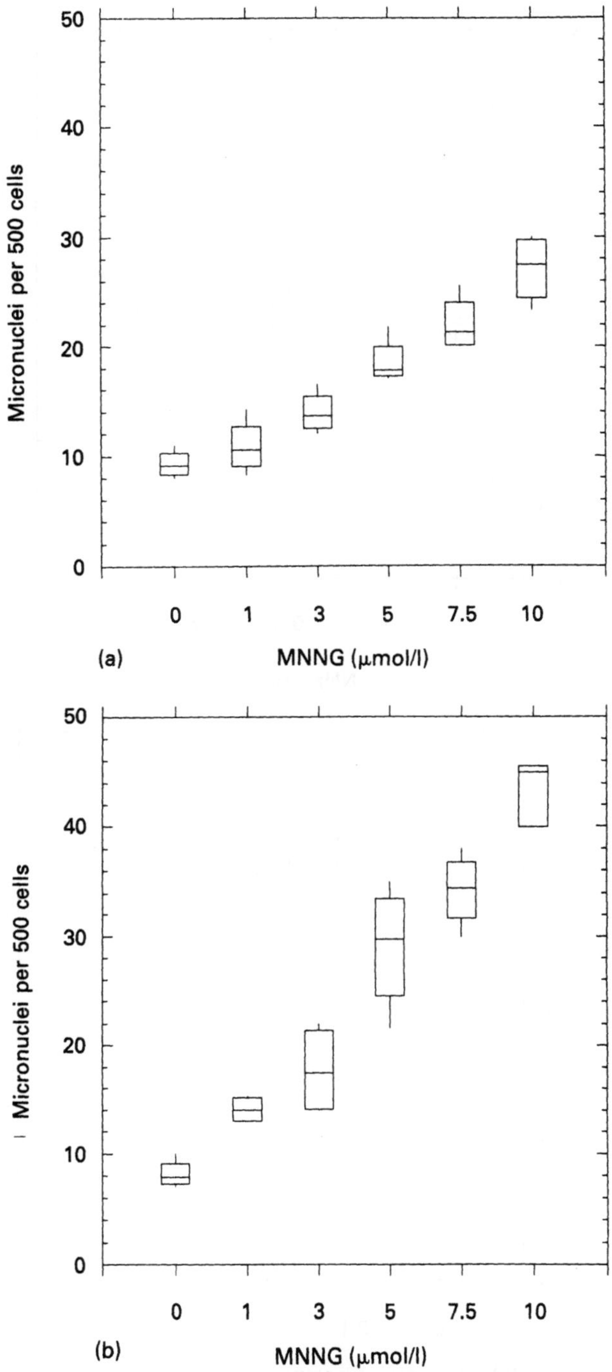

Figure 14.3 Box whisker plot of the induction of micronuclei by MNNG in (a) repair-proficient and (b) repair-deficient human fibroblast cultures.

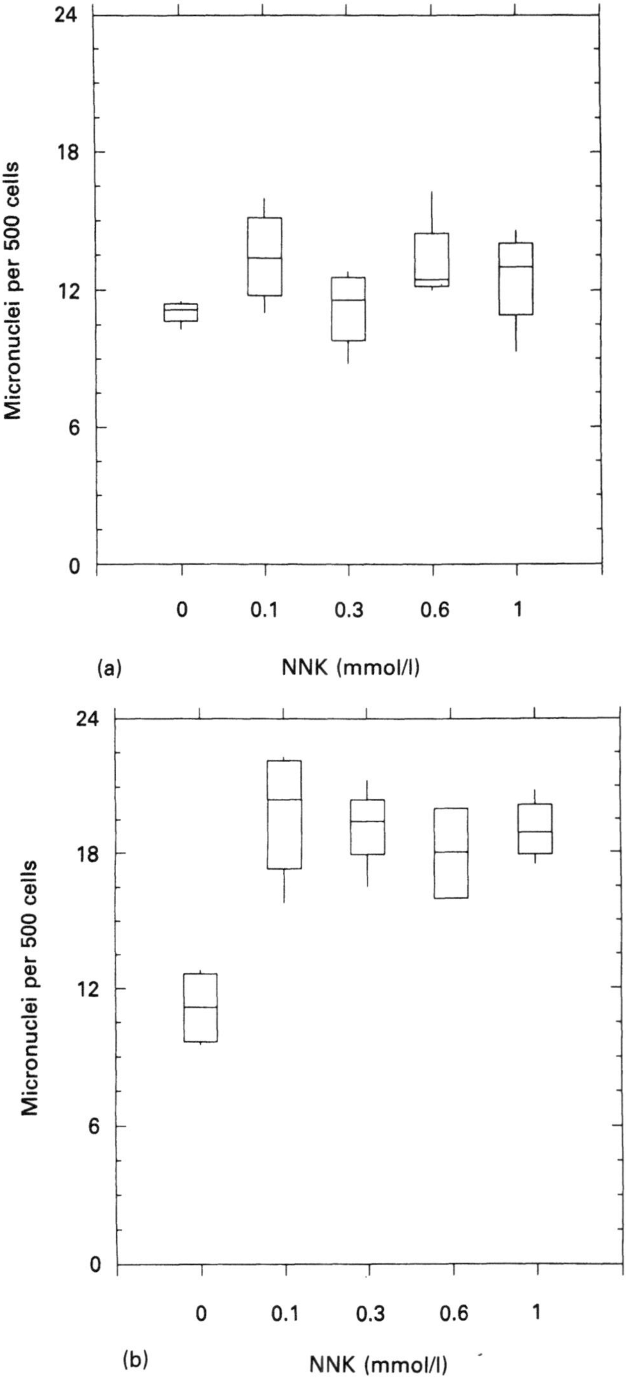

Figure 14.4 Box whisker plot of the induction of micronuclei by NNK in (a) repair-proficient and (b) repair-deficient human fibroblast cultures.

nitrosamine 4(metylnitrosamino)–1-(3-pyridyl)–1-butanone (NNK) as a DNA-methylating agent. In contrast to MNNG, NNK has to be metabolically activated. The ultimate DNA-methylating intermediate is probably methyldiazohydroxide (Hecht and Hoffmann, 1988). Fibroblast cultures with normal O^6-mgua-MT did not show an increased rate of micronuclei at all, even at very high concentrations of NNK (Fig. 14.4). In contrast, the four cell cultures with a low repair capacity exhibited a significantly enhanced micronucleus level at the lowest NNK concentration used (0.1 mmol/l), which could not be increased further by increasing the concentration of NNK. This plateau must be interpreted to result from a saturation of the monooxygenase involved in NNK oxidation at the alpha position (Hecht and Hoffmann, 1988). In fibroblasts with a normal activity of O^6-mgua-MT, the repair capacity is obviously sufficient to remove instantly almost every methyl group at the O^6-position of guanine prior to a DNA replication.

Thus, in contrast to the individual results with MNNG, even a qualitative difference is found between repair-deficient and repair-proficient individuals with NNK. The tobacco-specific nitrosamine NNK seems not to be genotoxic at all in repair-proficient cells. Any NNK-induced DNA damage seems to be exclusively dependent on the individual capacity to remove O^6-methylguanine.

Only a few experimental data correlate a broad range of a variable repair capacity with NNK sensitivity. A threshold of O^6-mgua-MT may exist, above which an individual is completely protected against DNA damage by NNK. Furthermore, no data show a relation between basal activity and inducible activity of O^6-mgua-MT, and nothing is known about the genetic inheritance of this trait in humans. This gap clearly needs to be filled before this parameter can be used to identify individuals sensitive to DNA-methylating agents.

14.5 CONCLUSIONS

O^6-mgua-MT is the first example of a DNA repair enzyme that determines in part the risk of developing a common type of cancer in humans. It has a proven variability in the human population that has been related to an individual susceptibility to lung cancer following exposure to DNA-methylating agents. Other mechanisms that determine the individual capacity to repair DNA alterations may contribute to a person's risk of developing other tumours as well. Malignant melanoma of the skin may be an example, because patients with a familial occurence of this tumour have an enhanced sensitivity to UV radiation (Roser *et al.*, 1989), which may well be explained by an impaired specific DNA repair system.

In general, an individual cancer risk from genotoxic agents results from a disturbed balance between the genotoxic burden and specific defence mechanisms. An unbalanced state may be due to impaired DNA repair as well as to a high genotoxic overloading.

REFERENCES

Babich, M. A. and Day III, R. S. (1987) Exogenous O^6-methylguanine inhibits adduct removal and sensitizes human cells to killing by the chemical carcinogen *N*-methyl-*N'*-nitro-*N*-nitrosoguanidine. *Mutat. Res.*, **184**, 245–54.

Belinsky, S. A., Walker, V. E., Maronpot, R. R. *et al.* (1987) Molecular dosimetry of DNA adduct formation and cell toxicity in rat nasal mucosa following exposure to the tobacco specific nitrosamine 4-(*N*-methyl-*N*-nitrosamino)–1-(3-pyridyl)–1-butanone and their relationship to induction of neoplasia. *Cancer Res.*, **47**, 6058–65.

Bignami, M., Terlizzese, M., Zijno, A. *et al.* (1987) Cytotoxicity, mutations and SCEs induced by methylating agents are reduced in CHO cells expressing an active mammalian O^6-methylguanine-DNA methyltransferase gene. *Carcinogenesis*, **8**, 1417–21.

Brent, T. P. (1986) Inactivation of purified human O^6-alkylguanine-DNA alkyltransferase by alkylating agents or alkylated DNA. *Cancer Res.*, **46**, 2320–3.

Cleaver, J. E. (1983) Xeroderma pigmentosum, in *The Metabolic Basis of Inherited Disease* (eds J. B. Stanbury, J. B. Wyngaarden, D. S. Fredrickson *et al.*), 5th edn, McGraw-Hill, New York, pp. 1227–48.

Day III, R. S., Babich, M. A., Yarosch, D. B. and Scudiero, D. A. (1987) The role of O^6-methylguanine in human cell killing, sister chromatid exchange induction and mutagenesis: A review. *J. Cell. Sci. Supp.*, **6**, 333–53.

Domoradzki, J., Pegg, A. E., Dolan, M. E. *et al.* (1984) Correlation between O^6-methylguanine-DNA-methyltransferase activity and resistance of human cells to the cytotoxic and mutagenic effect of *N*-methyl-*N*-nitro-*N*-nitrosoguanidine. *Carcinogenesis*, **5**, 1641–7.

Domoradzki, J., Pegg, A. E., Dolan, M. E. *et al.* (1985) Depletion of O^6-methylguanine-DNA-methyltransferase in human fibroblasts increases the mutagenic response to *N*-methyl-*N*-nitrosoguanidine. *Carcinogenesis*, **6**, 1823–6.

Doniger, J., Day III, R. S. and Dipaolo, J. A. (1985) Quantitative assessment of the role of O^6-methylguanine in the initiation of carcinogenesis by methylating agents. *Proc. Natl Acad. Sci. USA*, **82**, 421–5.

Elwood, J. M., Gallagher, R. P., Hill, G. B. and Pearson, J. C. G. (1985) Cutaneous melanoma in relation to intermittent and constant sun exposure – the western Canada melanoma study. *Int. J. Cancer*, **35**, 427–33.

Frosina, G. and Abbondandolo, A. (1985) The current evidence for an adaptive response to alkylating agents in mammalian cells, with special reference to experiments with in vitro cell cultures. *Mutat. Res.*, **154**, 85–100.

Hecht, S. S. and Hoffmann, D. (1988) Tobacco-specific nitrosamines, an important group of carcinogens in tobacco and tobacco smoke. *Carcinogenesis*, **9**, 875–84.

Kraemer, K. H., Coon, H. G., Petinga, R. A. *et al.* (1975) Genetic heterogeneity in xeroderma pigmentosum: complementation groups and their relationship to DNA repair rates. *Proc. Natl Acad. Sci. USA*, **72**, 59–63.

Kripke, M. L. (1979) Speculations on the role of ultraviolet radiation in the development of malignant melanoma. *J. Natl Cancer. Inst.*, **63**, 541–5.

Lijinsky, W. (1986) Mutagenesis, carcinogenesis, and alkylating properties of nitrosamines and related compounds, in *Genetic Toxicology of Environmental Chemicals. Part*

B: Genetic Effects and Applied Mutagenesis (eds C. Ramel, B. Lambert and J. Magnusson), Alan R. Liss, New York, pp. 141–8.

Lindahl, T. (1982) DNA repair enzymes. *Annu. Rev. Biochem.*, **51**, 61–87.

Lindahl, T. (1987) Regulation and deficiencies in DNA repair. *Br. J. Cancer*, **56**, 91–5.

Loveless, A. (1969) Possible relevance of O^6 alkylation of deoxyguanosine to the mutagenicity and cancerogenicity of nitrosamines and nitrosamides. *Nature*, **223**, 206–7.

Maher, V. M., Domoradzki, J., Corner, R. C. and McCormick, J. J. (1986) Correlation between O^6-alkylguanine-DNA alkyltransferase activity and resistance of human cells to the cytotoxic and mutagenic effects of methylating and ethylating agents, in *Biochemical and Molecular Epidemiology of Cancer* (ed. C. C. Harris), Allan R. Liss, New York, pp. 411–18.

Morimoto, K., Sato-Mizuno, M. and Koizumi, A. (1986) Adaptation-like response to the chemical induction of sister chromatid exchanges in human lymphocytes. *Hum. Genet.*, **73**, 81–5.

Painter, R. B. and Young, B. R. (1980) Radiosensitivity in ataxia-teleangiectasia: a new explanation. *Proc. Natl Acad. Sci. USA*, **77**, 7315–17.

Pegg, A. E. (1978) Dimethylnitrosamine inhibits exzymatic removal of O^6-methylguanine from DNA. *Nature*, **274**, 182–4.

Pegg, A. E. (1984) Methylation of the O^6 position of guanine in DNA is the most likely initiating event in carcinogenesis by methylating agents. *Cancer Invest.*, **2**, 223–31.

Roser, M., Böhm, A., Oldigs, M. *et al.* (1989) Ultraviolet-induced formation of micronuclei and sister chromatid exchange in cultured fibroblasts of patients with cutaneous malignant melanoma. *Cancer Genet. Cytogenet.*, **41**, 129–37.

Rüdiger, H. W., Schwartz, U., Serrand, E. *et al.* (1989) Reduced O^6-methylguanine repair in fibroblast cultures from patients with lung cancer. *Cancer Res.*, **49**, 5623–6.

Scudiero, D. A., Meyer, S. A., Clatterbuck, B. E. *et al.* (1984) Sensitivity of human cells strains having different abilities to repair O^6-methylguanine in DNA to inactivation by alkylating agents including chloroethylnitrosoureas. *Cancer Res.*, **44**, 2467–74.

Swift, J. E. (1976) Malignant disease in heterozygous carriers, in *Cancer and genetics*, (The National Foundation – March of Dimes Birth Defects: Original Article Series, Vol. XII, no. 1) (eds D. Bergsma, R. N. Schimke, R. L. Summitt *et al.*), Alan R. Liss, New York, pp. 133–44.

Trey, J. E. and Gerson, S. L. (1987) Enhanced induction of sister chromatid exchanges in resting vs. proliferating lymphocytes after inactivation of O^6-alkylguanine DNA alkyltransferase by O^6-methylguanine. *Proc. Annu. Meet. Am. Assoc. Cancer Res.*, **28**, 118.

Waldstein, E. A., Cao, E. -H. and Setlow, R. B. (1982) Adaptive resynthesis of O^6-methylguanine-accepting protein can explain the differences between mammalian cells proficient and deficient in methyl excision repair. *Proc. Natl Acad. Sci. USA*, **79**, 5117–21.

Willis, A. E. and Lindahl, T. (1987) DNA ligase I deficiency in Bloom's syndrome. *Nature*, **325**, 355–7.

Wynder, E. L. and Graham, E. A. (1985) Tobacco smoking as a possible etiologic factor in bronchiogenic carcinoma. *JAMA*, **253**, 2986–94.

Yarosh, D. B., Fornace, A. and Day III, R. S. (1985) Repair of O^6-methylguanine and methyl phosphotriesters in DNA by human O^6-alkylguanine alkyltransferase. *Carcinogenesis*, **6**, 949–53.

Yarosh, D. B., Rice, M. and Day III, R. S. (1984) O^6-methylguanine-DNA methyltransferase in human cells. *Mutat. Res.*, **131**, 27–36.

15 Interindividual variation in the level of DNA and protein adducts in humans

K. Hemminki

Department of Industrial Hygiene and Toxicology, Institute of Occupational Health, Helsinki, Finland, and Center for Nutrition and Topicology, Karolinska Institute, Huddinge, Sweden

15.1 SUMMARY

This chapter reviews the interindividual variations reported in DNA and protein binding in humans. In DNA binding, the apparent adduct levels vary to such an extent between studies that comparison is difficult. The only measurements whereby the method allows comparison of two populations, one from Sweden and the other from England, were those for hydroxyethylvaline concentrations in haemoglobin of smokers. The two smoker and control non-smoker populations showed identical results as far as absolute adduct levels and interindividual variation were concerned. The present survey suggests that when adjustment is made for the level of exposure in a defined population the interindividual variation in protein binding is usually less than 5-fold and in DNA binding less than 10-fold. However more carcinogens and new populations need to be studied before these results can be generalized.

15.2 INTRODUCTION

Binding of a chemical to DNA is thought to be the event which triggers the cascade of biochemical changes eventually leading to malignancy (Weinstein, 1988). This does not rule out the possibility that some chemicals may cause or contribute to cancer by mechanisms other than binding to DNA. In target tissues of experimental animals the levels of adducts correlate with tumour responses (Beland *et al.*, 1988; Wogan, 1988). This has not been directly studied in humans, but inferences can be made from animal studies and from the indirect data, discussed later, on groups at high risk of cancer. Because target tissues are usually not available in humans and the DNA binding studies resort to surrogate tissue, the above inference is less compelling. Generally, binding to protein, such as haemoglobin and albumin, correlates with binding to DNA and protein adducts can be used as a surrogate measure of DNA binding (Ehrenberg *et al.*, 1983; Neumann, 1984; Wogan, 1988).

The development of sensitive methods based on ^{32}P-postlabelling (Randerath *et al.*, 1981; Gupta *et al.*, 1982; Reddy and Randerath, 1986) or immunoassay (Poirier, 1981) have laid the groundwork for the determination of adducts in humans. The technology is new and constantly improving and expanding (Bartsch *et al.*, 1988). Due caution has to be exercised in the interpretation of the data obtained in terms of accuracy and identity of the measurements. Recently, interlaboratory comparisons have been carried out for the postlabelling technique, immunoassay and both, and reasonable correlations have been found (Hemminki *et al.*, 1988). Yet unknown variables exist and there is no proof that in human DNA the two assays measure identical adducts.

15.3 EXPOSURE TO CHEMICALS

Interindividual variation in the level of adducts can only be assessed if the absorbed doses of the chemicals, or at least their ambient air levels, are quantitated. This is rarely achieved, and extensive differences in the level of exposure are seen not only in the occupational environment, where personal exposure monitoring has a long tradition, but also in the general environment. Gradients of concentration spread from sources of pollution, and are known to be extensive, for example between urban and rural areas (WHO, 1987).

The US Environmental Protection Agency (EPA) has carried out an extensive monitoring programme on the total exposure of a selected number of organic volatile compounds in a few American communities (EPA, 1987). The thorough measurements demonstrate surprisingly large differences of exposure, ranging from 10- to 50-fold, depending on the compound, for the exposures of a single community. An example on the distribution of population is shown in Fig. 15.1, illustrating the proportion of population by exposure level. For air-borne particulate material even larger differences can be expected and fewer representative studies are available.

The requirement for quantitative exposure data narrows the scope of this review to a limited number of exposures, and even in such cases

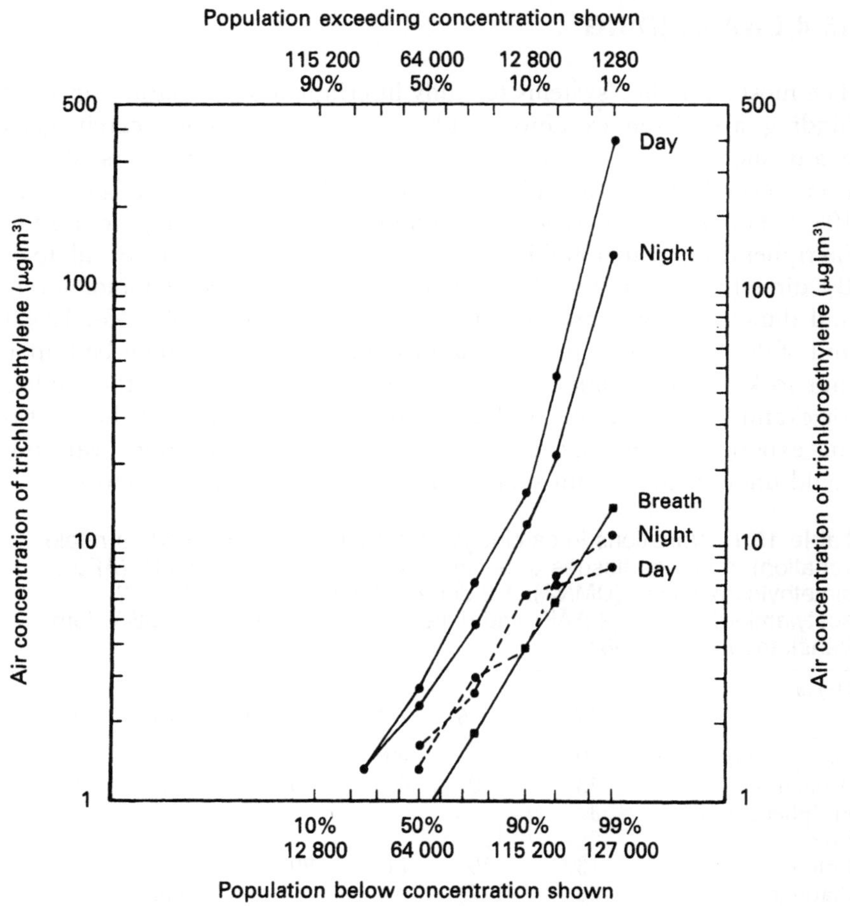

Figure 15.1 Distribution of population in a community in the USA with respect to air concentrations of trichloroethylene. Day and night samples were taken by personal monitors (•—•, n = 344), form exhaled air (■—■, n = 320) and by stationary outdoor monitors (•---•, n = 86) (from EPA, 1987).

uniformity of the absorbed dose needs to be questioned. For purposes other than comparison of the interindivial variation the above requirement is not required. The DNA- and protein-binding data may be used to corroborate the exposure data or to provide clues about the biologically effective agents among the complex mixtures.

15.4 DNA BINDING

The most simplified systems to study interindividual variations in DNA binding are tissue explants incubated with radioactive carcinogens. Many such studies were carried out by Harris and coworkers. Results from several studies are presented in Table 15.1 (Vähäkangas *et al.*, 1984). For benzo(a)pyrene the levels of DNA binding vary from 3-fold (peripheral lung) to 130-fold (colon). The tissue variations for aflatoxin B_1, dimethylnitrosamine, 1,2-dimethylhydrazine, 2-acetylaminofluorene and dimethylbenzanthracene are respectively 12–150-, 90–145-, 10–80- and 50-fold. Although these variations appear large, it would be important to know how uniformly the explants were treated for the studies, for example how necrotic and anaerobic the tissues were at the start of the experiment, the size of the tissue fragments, etc. Some variation could undoubtedly be introduced by such methodological factors.

Table 15.1 Variations in carcinogen-DNA binding (expressed as *n*-fold variation) in human tissues of benzo(a)pyrene (BP), aflatoxin B_1 (AFB), dimethylnitrosamine (DMNA), 1,2-dimethylhydrazine (1,2-DMH), 2-acetylaminofluorene (2-AAF) and dimethylbenzanthracene (DMBA) form Vähäkangas *et al.* 1984

Organ	*BP*	*AFB*	*DMNA*	*1,2-DMH*	*2-AAF*	*DMBA*
Oesophagus	99	70	90			
Bronchus	75	120	60	10	18	50
Peripheral lung	3					
Liver		12				
Colon	130	150	145	80		
Bladder	68	127			114	
Endometrium	70					

All the above compounds require metabolic activation, the enzymic activities of which are also known to vary extensively in tissue explants from different individuals (Vähäkangas *et al.*, 1984; Vähäkangas and Pelkonen, 1985, 1989). The inclusion in the above experiments of a directly reactive compound would have been useful to help resolve the contribution of differential metabolism from that of DNA binding.

15.4.1 Cisplatin

Cisplatin, a directly reactive anticancer agent, provides a simple illustration of the variation in the level of white blood cell DNA. Six patients receiving doses of between 20 and 100 mg/m^2 cisplatin by infusion were compared by Fichtinger-Schepman *et al.* (1987). In five patients the level of the main cisplatin DNA adduct, representing 65% of the total binding, varied less than 1.5-fold when the dose was taken into consideration. One patient (A) was unique as he developed 15 times more adduct than the three other patients receiving a similar dose. In the discussion the authors mention another person who developed 100 more adducts than patient A. In this small data set, where dosing was by infusion, five out of 6 individuals had uniform levels of adducts, and one out of six was an outlier (with a 15-fold higher level of adducts). There is a large set of data published on patients treated with cisplatin, but because the absolute adduct levels, determined by antibody, are much lower than those cited in the above study, it will not be discussed (Poirier *et al.*, 1985). The antibody was raised against platinated DNA and the immunoassay was carried out with total DNA.

15.4.2 Smokers

Smokers are another group in which interindividual variation may be estimated, and many studies have been carried out. However, the interpretation is complicated because many unidentified (aromatic) adducts are measured by ^{32}P-postlabelling assay or by antibodies raised against benzo(a)pyrene-modified DNA. Furthermore, most studies divide the groups into smokers and nonsmokers without further subdivision, and assay for non-target tissues.

Phillips *et al.* (1988) assayed coded non-tumourous lung samples by the ^{32}P-postlabelling technique and found a high correlation been the number of cigarettes smoked and the level of aromatic adducts. The data show, first, that considering all smokers as one group introduces a variation of a few fold, even though an average (male) smoker may consume about 20 cigarettes per day. Second, the data show about a 10-fold variation in the level of adducts for smokers of between 10 and 20 cigarettes per day. The remaining smoking levels are based on two or three individuals only and they exhibit a difference somewhat lower than two-fold. If the data are weighted by the number of individuals in each group the overall variation is 5-fold or slightly higher. In other postlabelling studies smoking-related adducts were noted in a number of tissues but no interindividual variation was reported (Everson *et al.*, 1986; Randerath *et al.*, 1989).

Table 15.2 summarizes the studies on smokers where smoking levels were not given in quantitative terms and non-target tissues were analysed. The interpretation of the interindividual variation is further complicated by the fact that in most assays some samples were negative, that is they were below the level of detection. Then the range of individual results is only given for positives, but this implies that the true difference is larger. The ranges in the level of adducts among smokers are from 3- to 16-fold. As the negatives were not considered (increasing

Table 15.2 DNA adducts in white blood cells of smokers (S) and nonsmokers (NS) (modified from Perera *et al.*, 1990)

Assay	*Population (n)*	*Range (μmol/mol)*[b]	*Positive (%)*	*Mean (positive)*[a] *(μmol/mol)*	*References*
PAH-DNA[c]	S (22)	0.054–0.17 (3×)	41	0.096	Perera *et al.* (1987)
(ELISA)[d]	NS (24)	0.035–0.14 (4×)	33	0.08	
PAH-DNA	S (32)	<0.032–0.29 (>9×)	9	0.192	Weston *et al.* (1989)
(USERIA)[e]	NS (49)	<0.032–0.13 (>4×)	4	0.096	
Aromatic DNA	S (4)	<0.002–0.014 (>7×)	100	0.0065	Phillips *et al.* (1986)
(^{32}P)[f]	NS (6)	0.004–0.01 (2.5×)	100	0.0065	

[a]The value given is the mean for positive samples i.e., those with detectable levels of adducts.
[b]μmol/mol: micromole adduct per mole DNA nucleotide.
[c]PAH-DNA: polyaromatic hydrocarbons-DNA.
[d]ELISA: enzyme linked immunosorbent assay.
[e]USERIA: ultrasensitive enzymatic radioimmunoassay.
[f]^{32}P: P-postlabelling.

the difference), but as smoking was undegraded according to the level (reducing the difference), it would be prudent to conclude that these studies suggest a difference of about 10-fold or less in DNA adducts of non-target tissues.

15.4.3 Occupational exposures

An example of published occupational studies is shown in Fig. 15.2, in which three laboratories assayed independently 53 samples from white blood cells of Finnish foundry workers by means of the ^{32}P-postlabelling assay (Savela *et al.*, 1989). The levels of aromatic adducts among the foundry workers, expressed per 10^8 nucleotides, ranged from 0.3 to 186 (620-fold) in laboratory 1, from 2.4 to 54 (23-fold) in laboratory 2 and from 0.04 to 129 (3200-fold) in laboratory 3. As the adduct levels in the control population had means of 1.7–3.1 per 10^8 nucleotides (only assayed for in two laboratories), some of the foundry worker samples were actually below the means of the controls, and the wide range in the adduct levels of foundry workers is weighted by these few low values. Because such low values may be 'false negatives', it might be prudent to summarize that in this large interlaboratory study, the levels of adduct levels among individual foundry workers ranged from the levels measured in the controls to those exceeding the controls 50- or 100-fold.

The contribution of differential exposure to this large variation was not assessed on an individual basis. However, when the individuals were

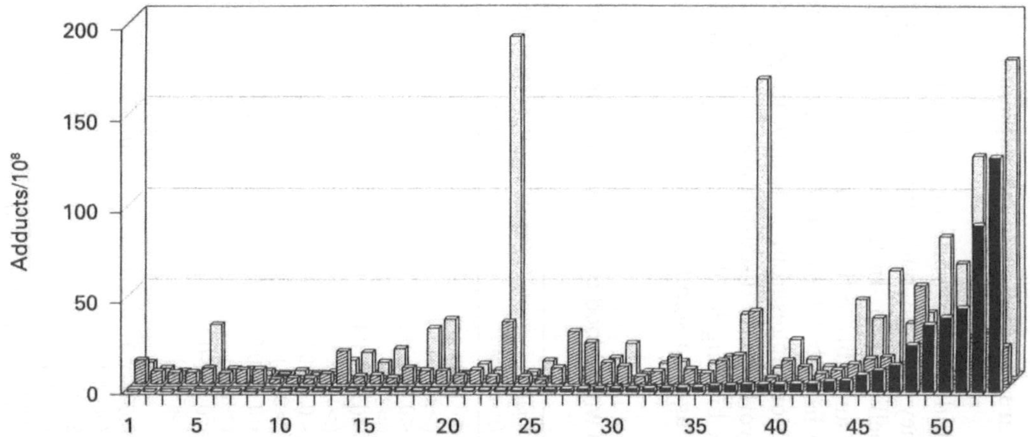

Figure 15.2 Aromatic adducts in white blood cell DNA from 53 foundry workers, assayed by three laboratories (differentiated by filled in, laboratory 1, hatched, laboratory 2, and shaded, laboratory 3, blocks) (from Savela *et al.*, 1989).

divided, according to job description, into three categories of exposure to benzo(a)pyrene (high: >0.2 $\mu g/m^3$, medium: 0.2–0.05 $\mu g/m^3$ and low: <0.05 $\mu g/m^3$) high correlations between adduct levels and assumed exposure were found in the postlabelling (Hemminki *et al.*, 1988) and immunological studies (Perera *et al.*, 1988). This indicates that the differences in adduct levels are, at least in part, explained by differential exposure.

Other occupational studies from white blood cell DNA are summarized in Table 15.3. They include both immunoassay (USERIA and ELISA) and ^{32}P-postlabelling data. In the immunoassay 35–86% of the samples were positive, and this inflates the apparent interindividual variations, which range from >3.4 to >137. In the postlabelling data the interindividual variation ranges from 50 to 550. In general these data agree with the foundry worker data discussed above. In the occupational setting, individual DNA binding levels for PAH-type aromatic adducts vary up to 100-fold, but again a significant contribution to the variation is the level of exposure.

Table 15.3 PAH-DNA in white blood cells of workers (modified from Perera *et al.*, 1990)

Assay	*Population (n)*	*Range (µmol/mol)*	*%Pos.*	*Mean (positive)*[a] *(µmol/mol)*	*Reference*
PAH-DNA (USERIA)	Coke oven (20)	<0.032–0.13 (>4×)	35	0.064	Shamsuddin *et al.*, 1985
PAH-DNA (ELISA)	Coke oven (27)	<0.032–10.9 (>3.4×)	66	2.27	Harris *et al.*, 1985
PAH-DNA (USERIA)	Coke oven (38)	<0.032–4.38 (>137×)	34	0.51	Haugen *et al.*, 1986
PAH-DNA (ELISA)	Foundry (35)	<0.01–0.90 (>90×)	86	0.20	Perera *et al.*, 1988
PAH-DNA (ELISA)	Coke oven (31)	<0.01–0.82 (>82×)	74	0.15	Hemminki *et al.* (1990)
Aromatic adducts (^{32}P)	Foundry[b] (25)	<0.002–0.1 (>50×)	100	0.018	Phillips *et al.*, 1988
Aromatic adducts (^{32}P)	Coke oven (63)	0.004–2.0 (× 500)	100	0.1–0.2	Hemminki *et al.* (1990)
		0.02–3.2 (× 160)	100	0.2	

[a]The value given is the mean for positive samples, i.e., those with detectable levels of adducts. Values in parentheses are means for all samples assayed.
[b]Medium and high exposures.
Abbreviations: see footnotes to Table 15.2.

15.4.4 DNA adducts in controls

Aromatic DNA adducts are found in the nonsmoking control individuals, who are not occupationally exposed to PAHs (see Table 15.2 for nonsmokers). With immunoassays, usually less than half of the samples are positive and show around 4-fold interindividual variation or more. According to the ^{32}P-postlabelling assays, all white blood cell DNA samples are usually positive at a level of <5 adducts/10^8 nucleotides (Savela *et al.*, 1989); 2.5-fold interindividual variation was seen in one study among nonsmokers (Table 15.2).

Two control populations were used in a postlabelling study of Polish coke workers: local controls living in communities around the coke plants (Fig. 15.3) and countryside controls (Fig. 15.4). The local controls showed an interindividual variation of 13-fold (laboratory 1) or 22-fold (laboratory 2) when two individuals (15 and 18) were not considered because they were likely to be occupationally exposed to PAHs. Much lower adduct levels were detected in workers from the countryside; the interindividual variation was 8.5- and 15-fold in respective laboratories 1 and 2 (Fig. 15.4). A likely explanation for this difference is environmental exposure to aromatic compounds in communities around the cokeries due to industrial sources of domestic coal-burning heating devices (Hemminki *et al.*, 1990).

It appears in general that the variation in the background level of adducts is less than that seen in the occupational settings. However, this may be artefactual because the measurements are close to the limit of detection and because the adducts measured may vary from one individual to another.

15.5 PROTEIN BINDING

Protein binding has mainly been measured in haemoglobin and serum albumin, different compounds showing preference for one or the other. For directly active compounds binding to albumin in particular is a more simple model than binding to white blood cell DNA, as no cellular barriers have to be crossed.

15.5.1 Cisplatin

Patients treated with various doses of cisplatin were sampled for haemoglobin-bound (Fig. 15.5) and total plasma-protein-bound (Fig. 15.6) platinum, determined by atomic absorption spectroscopy 0.3 h and over 10 weeks after the end of cisplatin infusion (Mustonen *et al.*, 1989). In the case of haemoglobin binding at 0–3 h after infusion, samples from the individuals within a factor of about 5. After 24 h the range appears slightly lower, between 3 and 4.

The binding to plasma proteins differed less (Fig. 15.6). Overall the

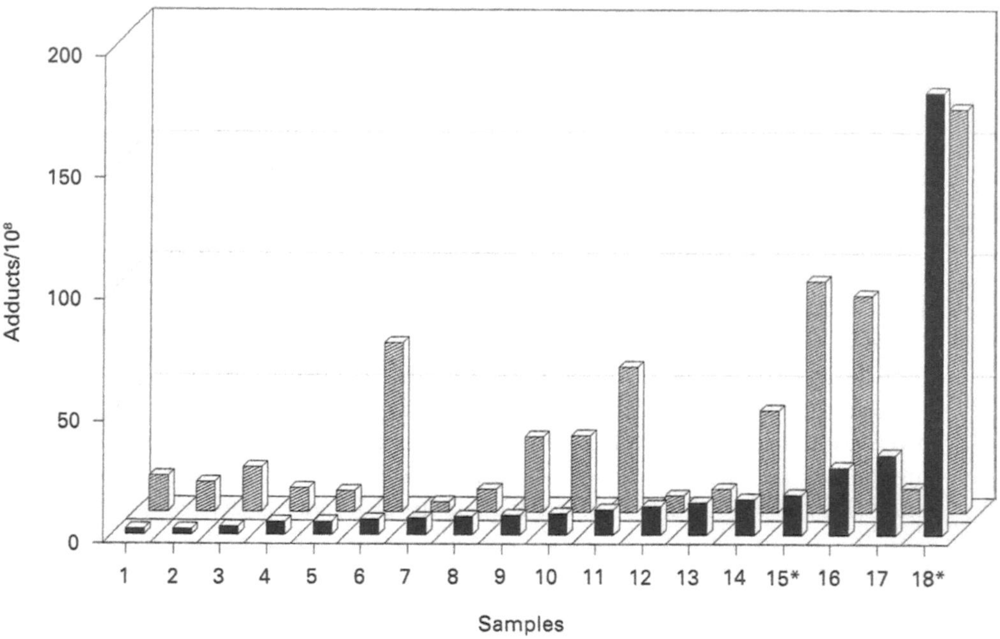

Figure 15.3 Postlabelling results from two laboratories (hatched, laboratory 1, and filled in blocks, laboratory 2) of white blood cell DNA from persons living close to cokeries; individuals 15 and 18 (asterixed) have occupations which may involve exposure to PAHs (Hemminki *et al.*, 1990).

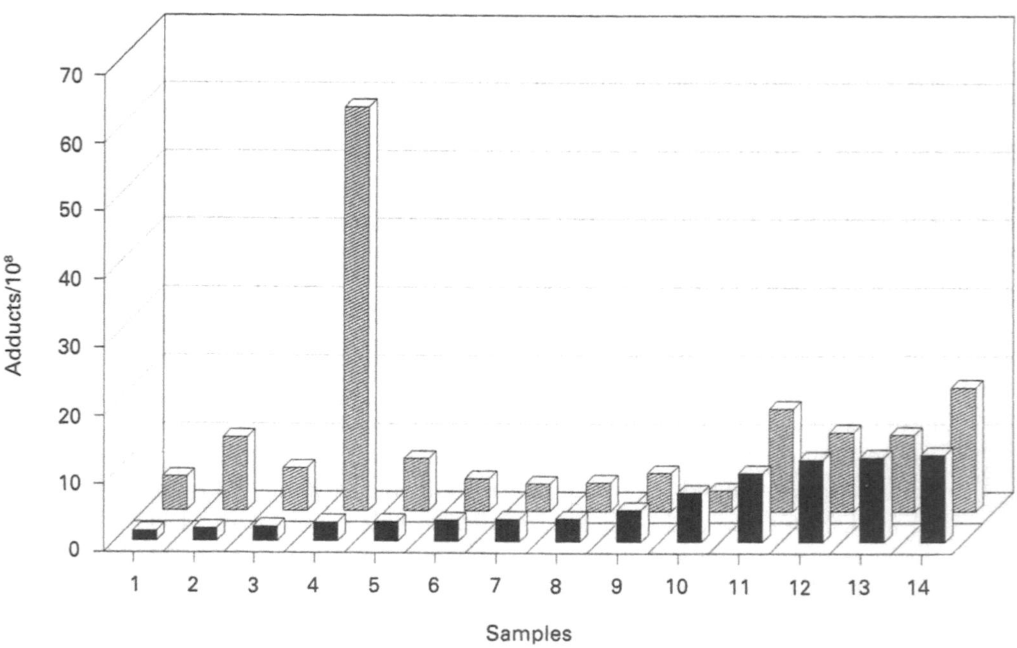

Figure 15.4 Postlabelling results, from the same laboratories as for Fig. 15.3, of countryside controls (from Hemminki *et al.*, 1990).

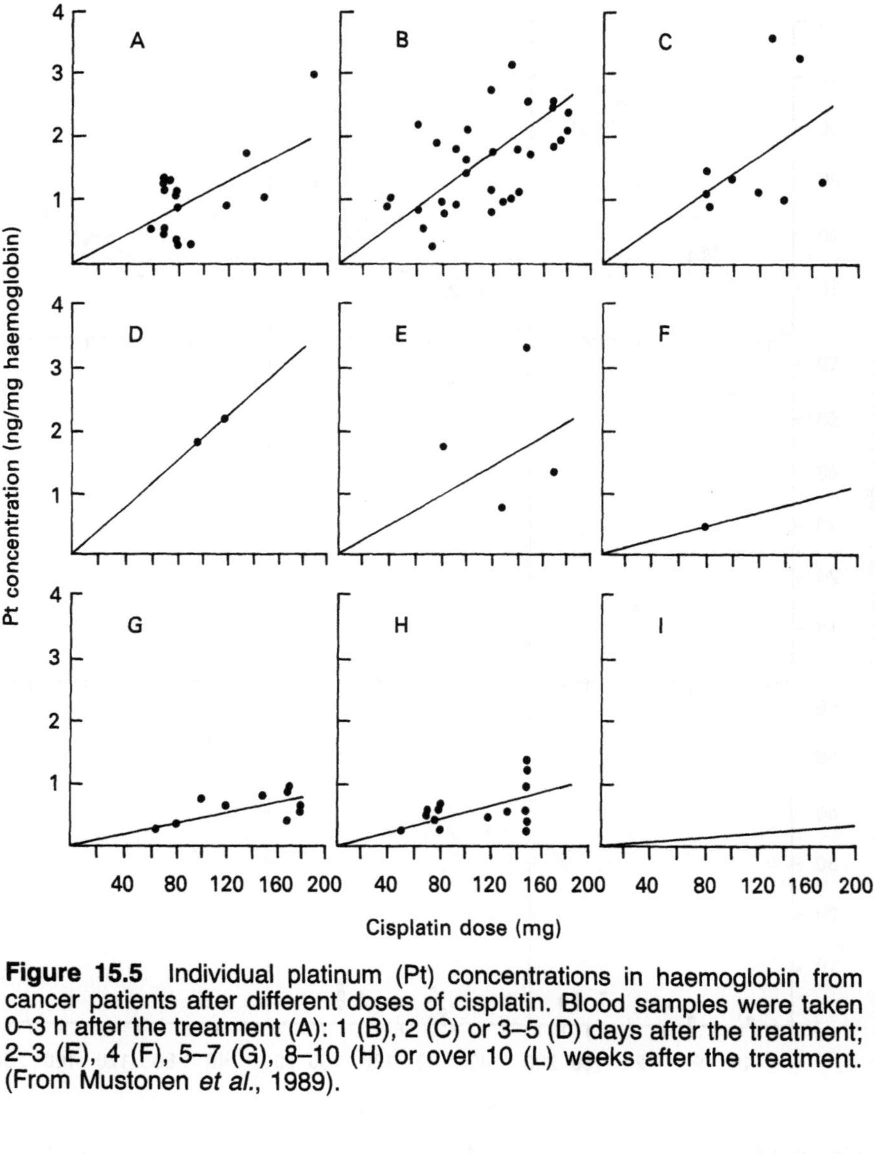

Figure 15.5 Individual platinum (Pt) concentrations in haemoglobin from cancer patients after different doses of cisplatin. Blood samples were taken 0–3 h after the treatment (A): 1 (B), 2 (C) or 3–5 (D) days after the treatment; 2–3 (E), 4 (F), 5–7 (G), 8–10 (H) or over 10 (L) weeks after the treatment. (From Mustonen *et al.*, 1989).

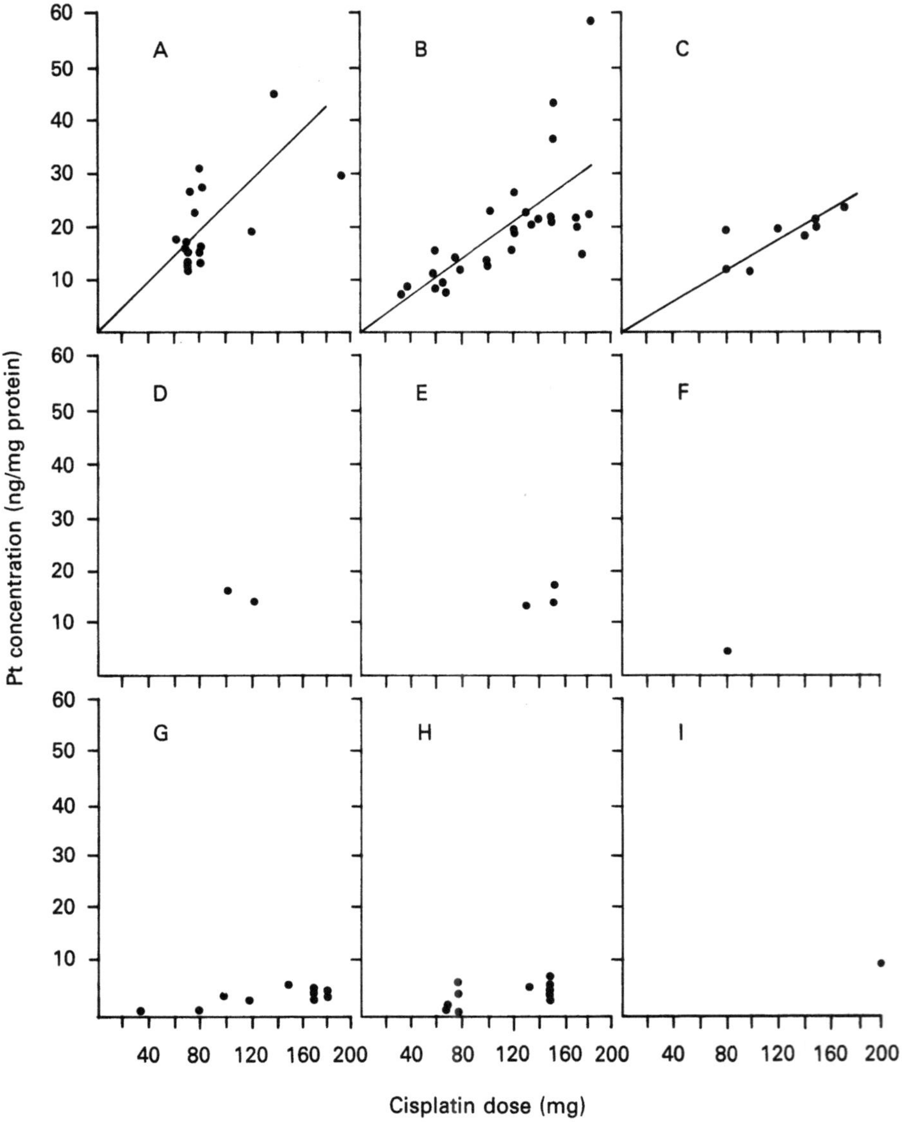

Figure 15.6 Individual platinum (Pt) concentration in plasma proteins from cancer patients after different doses of cisplatin. Blood samples were taken 0–3 h after the treatment (A); 1 (B), 2 (C) or 3–5 (D) days after the treatment; 2–3 (E), 4 (F) or 5–7 (G) weeks after the treatment. (From Mustonen *et al.*, 1989.)

binding levels showed about a 2.5-fold difference after 0–3 h and a 2-fold difference after 24 h.

15.5.2 Smokers

Tobacco smoke contains ethylene oxide and ethene, both of which can give rise to 2-hydroxyethylation of macromolecules. Such adducts were determined from *N*-terminal valine of haemoglobin in heavy smokers (>20 cigarettes/day) and 14 nonsmokers (Törnqvist *et al.*, 1986). The respective mean *N*-hydroxyethylvaline levels were 389 and 58 pmol/g haemoglobin, respectively. The ranges were 3-fold among smokers and 4-fold among nonsmokers. After recalibration, which slightly decreased the mean values, the respective interindividual differences were 4-fold for smokers and 2.5-fold for nonsmokers (Törnqvist, 1989). Some variation in hydroxyethylation may be explained by differential smoking habits. The hydroxyethylations among nonsmokers may originate from many sources such as lipid peroxidation (Törnqvist, 1989).

In the study conducted by Bailey *et al.* (1988), the range of hydroxyethylvaline in haemoglobin in 23 nonsmokers was 5-fold. Among smokers, different smoking levels were considered; summing the data over all smoking levels, the difference in hydroxyethylation in 26 smokers was about 3-fold.

Tobacco smoke also contains several aromatic amines and it was noted by Bryant *et al.* (1987) that the levels of 4-aminobiphenyl correlate with smoking, that is smokers and nonsmokers showed respective levels of 154 and 28 pg 4-aminobiphenyl/g haemoglobin. 4-Aminobiphenyl and other aromatic amines are metabolized mainly in the liver to the corresponding hydroxylamines; these are oxidated by oxyhaemoglobin in red blood cells and converted to *N*-nitroso compounds capable of binding to haemoglobin (Bryant *et al.*, 1987).

In another study of smokers, based on the same individuals as the above study, three blood samples were drawn from smokers of at least 20 cigarettes per day (mean 28) in one-day intervals; two samples were drawn from nonsmokers (Perera *et al.*, 1987). The results for 4-aminobiphenyl were given for two samples in each group; the mean levels for smokers and nonsmokers were respectively 139–155 and 32–36 pg/g. The range of values among smokers was 3.4-fold (sample 1) and 1.9-fold (sample 3); as sample 1 contained twice as many samples, the weighted range in individual values was about 3. However, some of this difference is likely to be explained by differential smoking habbits. The overall range of haemoglobin-bound 4-aminobiphenyl residues among nonsmokers was about 6. The source of background levels of 4-aminobiphenyl is not known.

15.6 CONCLUSIONS

Macromolecular adducts have been recently introduced as dosimeters of covalent binding. The techniques used for the measurement of DNA and protein binding are different. Measurement of protein binding products is by standard methods of analytical chemistry which display high precision. The methods for DNA binding are complex and usually do not measure an identified chemical. Thus some apparent variation between individuals may be introduced by analytical inaccuracy. Hence it is useful to summarize the results of protein binding first, because the results reflect true interindividual variability.

Table 15.4 Summary of the interindividual variation noted in the protein binding studies discussed

Exposure	*Variation noted*	*Comments*
Cisplatin, haemoglobin	3–5	Directly reactive compound
Cisplatin, plasma protein	2–2.5	
Smoking, hydroyethylvaline in haemoglobin	3–4	Source may be ethene or ethylene oxide
Smoking, 4-aminobiphenyl in haemoglobin	3	Smoking at least 20 cigarettes a day
Background, hydroyethylvaline in haemoglobin	2.5–5	Source is unknown
Background, 4-aminobiphenyl in haemoglobin	6	Source is unknown

The protein binding results for cisplatin and smoking-related exposures vary only 2–5-fold between individuals (Table 15.4). While cisplatin is a directly acting compound, most hydroxyethylations and 4-aminobiphenyl binding require metabolic activation. Thus the small variation, 3–4-fold, in the smoking-related adducts is particularly significant. Moreover, two of the three studies did not absolutely adjust for the level of smoking, making 3–4 an overestimate of the true variation. Background hydroxyethylation and 4-aminobiphenyl binding varied 2.5–6-fold between individuals, but the source of these adducts is unknown.

The results on DNA binding are summarized in Table 15.5. The few patients studied for cisplatin binding in white cell DNA displayed small interindividual differences (1.5-fold) although outliers may exist. Clearly more data are needed for conclusions to be drawn. Interestingly, pharmacokinetic and metabolic differences appear generally much larger among individuals and there may be a correlation with the risk of cancer (Caporaso *et al.*, 1989; Vähäkangas and Pelkonen, 1989).

Table 15.5 Summary of the interindividual variation noted in the DNA binding studies discussed

Exposure/system	*Variation noted*	*Comments*
Many carcinogens in vitro	10–150	Methodic variation not ruled out
Cisplatin	1.5(15)[a]	Few patients, outliers may exist
Smoking, aromatic postlabelling	5–10	Lung tissue
Smoking, aromatic, PAH Postlabelling and immunoassay	10	Mainly white blood cells
Foundries, cokeries, postlabelling and immunoassay	100	Part of variation explained by intensity of exposure
Background, aromatic	2–20	Indentity of adducts may vary; close to limit of detection

[a]One individual.

Smoking-related aromatic adducts showed an interindividual variation of 5–10-fold in lung tissue and 10-fold in white blood cells and placental samples. In an occupational setting variations of about 100-fold have been noted for aromatic adducts between individuals, but differences in the level of exposure explain this to some extent. Background aromatic adducts have shown interindividual variation of 2 to 20, but the identity of the adducts may vary between the individuals, invalidating such comparisons.

The present study surveyed essentially the interindividual variations in single studies, usually consisting of homogeneous populations. In the different DNA binding studies the apparent adduct levels between studies vary to such an extent that comparison of the studies is difficult. The only measurements where the method allows comparison of two populations, one from Sweden and the other from England, were those for hydroxyethylvaline concentrations in the haemoglobin of smokers. The smoker and control nonsmoker populations showed identical results as far as absolute adduct levels and interindividual variation were concerned. The present survey suggests that when adjustment is made for the level of exposure in a defined population, the interindividual variation in protein binding is usually less than 5-fold and in DNA binding less than 10-fold. However, more carcinogens and new populations need to be studied before these results can be generalized.

REFERENCES

Bailey, E., Brooks, A. G. F., Dollery, C. T. *et al.*, (1988) Hydroxyethylvaline adduct formation in haemoglobin as a biological monitor of cigarette smoke intake. *Arch. Toxicol.*, **62** 247–53.

Bartsch, H., Hemminki, K. and O'Neill, I. K. (eds) (1988) *Methods for Detecting DNA Damaging Agents in Humans: Applications in Cancer Epidemiology and Prevention.* IARC Scientific Publication No. 89, International Agency for Research on Cancer, Lyon.

Beland, F. A., Fullerton, N. F. Kinouchi, T. and Poirier, M. C. (1988) DNA adducts formation during continuous feeding of 2-acetylaminofluorene at multiple concentrations, in *Methods for Detecting DNA Damaging Agents in Humans: Applications in Cancer Epidemiology and Prevention* (eds H. Bartsch, K. Hemminki and I. K. O'Neill), IARC Scientific Publication No. 89, International Agency for Research on Cancer, Lyon, pp. 175–80.

Bryant, M. S., Skipper, P. L., Tannenbaum, S. R. and Maclure, M. (1987) Hemoglobin adducts of 4-aminobiphenyl in smokers and nonsmokers. *Cancer Res.*, **47**, 602–8.

Caporaso, N., Hayes, R. B., Dosemeci, M. *et al.* (1989) Lung cancer risk, occupational exposure, and the deprisoquine metabolic phenotype. *Cancer Res.*, **49**, 3675–9.

Ehrenberg, L., Moustacchi, E. and Osterman-Golkar, S. (1983) Dosimetry of genotoxic agents and dose-response relationships in their effects. *Mutat. Res.*, **123**, 121–82.

EPA (1987) *The Total Exposure Assessment Methodology (TEAM) Study: Summary and Analysis*, Vol. 1, US Environmental Protection Agency, Washington DC.

Everson, R. B., Randerath, E., Santella, R. M. *et al.*, (1986) Detection of smoking-related covalent DNA Adducts in human placenta. *Science*, **231**, 54–7.

Fichtinger-Schepman, A. M. J., van Oosterom, A. T., Lohman, P. H. M. and Berends, F. (1987) Cis-diaminedichloroplatinum (II)-induced DNA adducts in peripheral leukocytes from seven cancer patients: quantitative immunochemical detection of the adduct induction and removal after a single dose of cis-diaminedichloroplatinum (II). *Cancer Res.*, **47**, 3000–4.

Gupta, R. C., Reddy, M. V. and Randerath, K. (1982) ^{32}P-postlabelling analysis of non-radioactive aromatic carcinogen-DNA adducts. *Carcinogenesis*, **3**, 1081–92.

Harris, C. C., Vähäkangas, K., Newman, M. J. *et al.*, (1985) Detection of benzo(a)pyrene diol epoxide-DNA adducts in peripheral blood lymphocytes and antibodies to the adducts in serum from coke oven workers. *Proc. Natl Acad. Sci. USA.*, **82**, 6672–6.

Haugen, A., Becher, G., Benestad, C. *et al.* (1986) Determination of polycyclic aromatic hydrocarbons in the urine, benzo(a)pyrene diol epoxide DNA adducts in lymphocyte DNA, and antibodies to the adducts in sera from coke oven workers exposed to measured amounts of polycyclic aromatic hydrocarbons in the work atmosphere. *Cancer Res.*, **46**, 4178–83.

Hemminki, K., Grzybowska, E., Chorazy, M. *et al.*, (1990) *DNA Adducts in Humans Related to Occupational and Environmental Exposure to Aromatic Compounds*, (eds H. Vainio, M. Sorsa and A. J. McMichael), International Agency for Research on Cancer, Lyon. IARC Scientific Publications No. 104, pp. 81–92

Hemminki, K., Perera, F. P., Phillips, D. H. *et al.*, (1988) Aromatic DNA adducts in white blood cells of foundry workers, in *Methods for Detecting DNA Damaging Agents in Humans: Applications in Cancer Epidemiology and Prevention*, (eds H. Bartsch,

K. Hemminki and I. O'Neill), IARC Scientific Publication No. 89, International Agency for Research On Cancer, Lyon, pp. 190–5.

Mustonen, R., Hietanen, P., Leppälä, S. *et al.* (1989) Determination of cis-diaminedichloroplatinum (II) in plasma proteins and hemoglobin of cancer patients. *Arch. Toxicol.*, **63**, 361–6.

Neumann, H.-G. (1984) Analysis of hemoglobin as a dose monitor for alkylating and arylating agents. *Arch. Toxicol.*, **56**, 1–6.

Perera, F. P., Hemminki, K., Young, T. L. *et al.* (1988) Detection of polycyclic aromatic hydrocarbon-DNA adducts in white blood cells of foundry workers. *Cancer Res.*, **48**, 2288–91.

Perera, F., Mayer, J., Santella, R. M. *et al.* (1990) Biologic markers in risk assessment for environmental carcinogens. *Environ. Health Perspect.*, in press.

Perera, F. P., Santella, R. M., Brenner, D. *et al.* (1987) DNA adducts, protein adducts and sister chromatid exchange in cigarette smokers and nonsmokers. *J. Natl Cancer Inst.*, **79**, 449–56.

Phillips, D. H., Hemminki, K., Alhonen, A. *et al.* (1988) Monitoring occupational exposure to carcinogens: detection by ^{32}P-postlabelling of aromatic DNA adducts in white blood cells from iron foundry workers. *Mutation Res.*, **204**, 531–41.

Phillips, D. H., Hewer, A. and Grover, P. L. (1986) Aromatic DNA adducts in human bone marrow and peripheral blood leukocytes. *Carcinogenesis*, **7**, 2071–5.

Phillips, D. H., Hewer, A., Martin, C. N. *et al.* (1988) Correlation of DNA adducts levels in human lung with cigarette smoking. *Nature*, **336**, 790–2.

Poirier, M. C. (1981) Antibodies to carcinogens – DNA adducts. *J. Natl Cancer Inst.*, **67**, 515–19.

Poirier, M. C., Reed, E., Zwelling, L. A. *et al.* (1985) Polyclonal antibodies to quantitate cis-diaminedichloroplatinum(II)-DNA adducts in cancer patients and animal models, *Environ. Health Perspect.*, **62**, 89–94.

Randerath, E., Miller, R. H., Mittal, D. *et al.* (1989) Covalent DNA damage in tissues of cigarette smokers as determined by ^{32}P-postlabelling assay. *J. Natl Cancer Inst.*, **81**, 341–7.

Randerath, K., Reddy, M. V. and Gupta, R. C. (1981) ^{32}P-labelling test for DNA damage. *Proc. Natl Acad. Sci. USA.* **78**, 6126–9.

Reddy, M. V. and Randerath, K. (1986) Nuclease P_1-mediated enhancement of sensitivity of ^{32}P-postlabelling test for structurally diverse DNA adducts. *Carcinogenesis*, **7**, 1543–51.

Savela, K., Hemminki, K., Hewer, A. *et al.* (1989) Interlaboratory comparison of the ^{32}P-postlabelling assay for aromatic DNA adducts in white blood cells of iron foundry workers, *Mutation Res.*, **224**, 485–92.

Shamsuddin, A. K. M., Sinopoli, N. T., Hemminki, K. *et al.* (1985) Detection of benzo(a)pyrene:DNA adducts in human white blood cells. *Cancer Res.*, **45**, 66–8.

Törnqvist, M. (1989) *Monitoring and Cancer Risk Assessment of Carcinogens Particularly Alkenes in Urban Air*, Univeristy of Stockholm, Stockholm.

Törnqvist, M., Österman-Golkar, S., Kautiainen, A. *et al.* (1986) Tissue doses of ethylene oxide in cigarette smokers determined from adduct levels in hemoglobin. *Carcinogenesis*, **7**, 1519–21.

Vähäkangas, K., Antrup, H. and Harris, CC. (1984) Interindividual variation in carcinogen metabolism, DNA damage and DNA repair, in *Monitoring Human Exposure to Carcinogenic and Mutagenic Agents* (eds. A. Berlin, M. Draper, K. Hemminki and H. Vainio), IARC Scientific Publication No. 59, International Agency for Research on Cancer, Lyon, pp. 85–98.

Vähäkangas, K. and Pelkonen, O. (1985) Host variations in carcinogen metabolism

and DNA repair, in *Familian Cancer* (ed. W. Müller), S. Karger, Basel, pp. 238–41.
Vähäkangas, K. and Pelkonen, O. (1989) Host variations in carcinogen metabolism and DNA repair, in *Genetic Epidemiology of Cancer* (eds H. T. Lynch and T. Hirayama), Wolfe Medical Publishers, London.
Weinstein, I. B. (1988) The origins of human cancer: molecular mechanisms of carcinogenesis and their implications for cancer prevention and treatment. *Cancer Res.*, **48**, 4135–43.
Weston, A., Rowe, M. L., Manchester, D. K. *et al.* (1989) Fluorescence and mass spectral evidence for the formation of benzo(a)pyrene anti-diol-epoxine-DNA and -hemoglobin adducts in humans. *Carcinogenesis*, **10**, 251–7.
WHO (1987) *Air Quality Guidelines for Europe*, World Health Organization, Copenhagen.
Wogan, G. N. (1988) Detection of DNA damage in studies on cancer etiology and prevention, in *Methods for Detecting DNA Damaging Agents in Humans: Applications in Cancer Epidemiology and Prevention* (eds H. Bartsch, K. Hemminki and I. K. O'Neill), IARC Scientific Publication No. 89, International Agency for Research on Cancer, Lyon, pp. 32–51.

PART SIX

Ethical and Regulatory Aspects

16 Ethical aspects of genetic predisposition to disease

P. Grandjean

Department of Environmental Medicine, Odense University, Odense, Denmark

16.1 SUMMARY

The fact that some individuals are much more susceptible to disease caused by chemical exposure will raise important ethical questions. Conflicts may occur in relation to moral principles, such as fairness, equity, right to privacy and the Hippocratic rules of nonmaleficence and beneficence. However, such moral obligations do not provide detailed guidance, because they sometimes result in conflicts. Also, utilitarian principles do not constitute a complete moral framework that would be widely acceptable. With genetic predisposition, the most serious problems relate to potential public interest in paternalistic measures, equal rights in free enterprise, equal right and opportunity to find work, fairness towards the handicapped, individual freedom and right to privacy. In addition, these concerns cause a need for extensive scientific documentation. Also, ways to deal with uncertainties must be considered. In a pluralistic society, these questions cannot be resolved by scientists alone or by any other group in isolation. Whether or not the ultimate goal can be considered a social contract, ethical conflicts can only be resolved in an open, informed forum. A public discourse on these ethical concerns would therefore seem highly pertinent.

16.2 BACKGROUND

The availability of increasing numbers of DNA probes, the polymerase chain reaction and other technological advances have greatly improved the prospects for large-scale genetic screening (Gibbs and Caskey, 1989). With the growth of information in pharmocogenetics and ecogenetics, tests are becoming available to determine individual predisposition to toxic effects of chemicals. In 1982, the US Office of Technology Assessment (OTA) found that limited genetic screening was being performed at a small number of industrial facilities in the USA. However, recent contact with key persons in several countries failed to identify cases of current genetic testing in the workplace. A new survey is now being undertaken by the OTA.

New techniques involving chromosomes and DNA are frequently regarded a threat to individual privacy and freedom. Thus, release of information in the news media often triggers emotional reactions. For example, a study was recently proposed to screen for chromosomal abnormalities in a cohort of Danish school children. The mere suggestion that genetic factors may be important for performance in school resulted in public protests and subsequent intervention by the Minister of Health. The critique mainly related to the rights to privacy and the risks of discrimination. Little attention was paid to the possible benefits, such as early identification of children in need of special help in school. When the plans were cancelled by the Minister, the discussion of ethical issues died out. In a longer-term perspective, such case-by-case solutions are likely to become contradictory. As no principles have been laid down for decision making, the confusion will tend to propagate itself.

Open discussion of the basic issues would be useful to clarify the empirical aspects and identify where moral questions are involved. However, a rational discourse on moral values may not necessarily lead to a resolution of the disagreement. No standard answers are uniformly acceptable and no solution is necessarily 'right'. Therefore, there are no experts who can always solve ethical problems by referring to truth.

The words 'ethical' and 'moral' are often used interchangeably. In the following, 'ethical' will refer to overall concerns, while 'moral' will relate to individual obligations. Thus, this paper will aim at identifying major ethical issues, rather than trying to resolve conflicting moral views concerning genetic predisposition to disease. This discussion is based on recent treatises on ethical concerns in relation to medicine (Veatch, 1981) and occupational health (Lappé, 1983; Ashford *et al.*, 1984; Rest and Patterson, 1986).

16.3 MORAL RULES

Certain moral rules are needed to provide a framework for ethical considerations. Moral obligations can be presented as deontological imperatives. Thus, they are generally based upon the notion of duty; an action is needed because it is inherently 'good'. Sometimes referred to as the golden rule, the law of moral reciprocity (or mutuality) requires that you do to others as you would have others do to you. Specific moral duties can then be deduced, taking into account, for example, whether or not the receiver is cognizant of the action, or the action is solicited. Because of some inherent difficulties in interpreting extensions of this basic duty, the 'principle of generic consistency' has been proposed: Act in accordance with generic rights (i.e. to freedom and well-being) of your recipients as well as of yourself (Gewirth, 1986).

The Hippocratic rules of nonmaleficence and beneficence can be looked upon as extensions of the basic moral principle. These moral obligations would appear particularly relevant in connection with screening for genetic predisposition to disease. However, although simple and seemingly logical, they offer no simple solutions in a world with a high degree of specialization, competition for limited resources and resulting pluralism and conflicts of interest. Thus, the Hippocratic rules give little guidance how to protect hypersusceptible individuals (beneficence) without violating other moral duties.

Another moral principle is fairness: like persons must be treated alike. Again, this moral law appears simple, but its practical use is complex (Lappé, 1983). For example, to which degree does fairness indicate that genetically susceptible individuals deserve protection or treatment that may appear preferential? And who owes them the protection – the employer or society?

Equity is a related moral obligation that no individual should be treated unequally unless he or she differs in a feature that is proven to affect his or her ability to perform an assigned job function, or unless he or she differs in willingness to accept the risks involved (Lappé, 1983). Again, interpretation of the egalitarian principle is not simple. Thus, genetic predisposition to disease means that humans are not equal in their resistance to environmental exposures; but what kinds of differential treatment are then necessary to restore equity?

In discussions on moral duties, reference will often be made to basic rights, as reflected in national constitutions or international conventions. However, ethical concerns are not only a matter of interpreting the law. The ultimate goal is to find solutions that are legitimate. Also, although we often tend to focus on rights alone, rights and responsibilities belong together. A balance is necessary and would need a cognizant approval of those affected by it.

In some situations, different moral duties appear to result in conflict. These disagreements are difficult or impossible to resolve, because 'good' actions cannot be viewed in isolation. Thus, the deontological principles obviously offer only a partial solution to ethical problems.

16.4 UTILITARIANISM

An attractive alternative is to consider benefits that are not limited to the individual patient with whom the physician or geneticist is interacting. Thus, in the utilitarian approach, a solution is sought that will bring the greatest benefit to the greatest number of people.

Resources in society are limited. As the health care system is facing severe budgetary restraints, we need to consider to which degree funds should be allocated to pay for new efforts in relation to the benefits expected. In this regard, screening for genetic predisposition to disease could seem a very attractive possibility.

However, utilitarianism would, in principle, allow procedures which do not result in a net benefit to the patient or which even produce considerable harm. Only the sum of satisfaction matters, but not the way in which this sum is distributed among individuals and over time. In following strict utilitarian principles, risks and benefits could be redistributed, thus allowing that expenses for the effort be passed to individuals other than those who enjoy the benefits. In accordance with utilitarian principles, a Swedish physician argued some years ago that the interests of society should overrule the wishes of a particular patient who had glaucoma and refused an operation. Without an operation, the patient would inevitably become blind and a financial burden to society. Thus, the physician argued that the operation should be forced upon the patient. Similarly, utilitarian considerations would be likely to suggest widespread use of pre-employment genetic screening to detect inherited susceptibility and application of regulations that would severely discriminate against certain ethnic groups.

An additional limitation of utilitarianism relates to the uncertainties and the difficulties in comparing different kinds of benefits and in comparing them to various types of costs and disadvantages. Thus, we probably never know enough to evaluate all relevant aspects of a particular effort, such as screening for genetic susceptibility. Even if we could get the right data, how do we compare various types of, for example, incapacitation? Also, as some benefits are projected to occur in a distant future, the comparison of risks and benefits must therefore also take the time scale into account.

Although powerful, utilitarianism obviously does not constitute a complete framework of moral rules that could receive wide acceptance. However, in some situations it may provide useful guidance in conjunction with deontological principles, such as the moral requirement of nonmaleficence. Another possible supplement to the utilitarian approach is the so-called distributive justice, that the benefits should generally accrue to those who bear the cost.

16.5 THE ETHICAL DISCOURSE

In modern society, agreement cannot be expected concerning moral obligations, religious duties and legal implications. Based on democratic principles and mutual tolerance, a pluralistic society will have to deal with a range of opinions, and ways must be found to obtain consensus on the resolution of conflicts. No single decision is necessarily more 'ethical' than another. Thus, ethics can be looked upon as part of our culture and political life. A basic premise is that all parties concerned have a right to contribute to the resolution of ethical conflicts.

The physician or geneticist is a key person in this process, especially because a detailed review of the scientific information is necessary as a starting point for the ethical considerations. However, health professionals may not necessarily be aware of related ethical aspects of the applications of medical genetics. Also, the very fact that he or she considers the profession important enough to invest years of study and long working hours may well lead to a professional bias.

As an ultimate goal for the ethical discourse, a social contract could be envisaged that would govern the application of screening for genetic predisposition. Rawls (1971) developed the idea of social justice and fairness as a basis for such a contract. The rules should be in accordance with a so-called original position. They should be acceptable to a group of people who 'under a veil of ignorance' were to establish the practices and organization of society of which they were to become members, their places in society not yet known.

A key principle would be that each individual would have an equal right to the most extensive liberty compatible with similar liberty for others (Rawls, 1971). In other words, tests for genetic susceptibility should provide health benefits without compromising values considered equal or higher, for example autonomy, justice or respect for persons (Lappé 1983). More specifically, when examining the possibilities, the following questions should be considered (Lappé 1983). (1) Does the program use the least restrictive of several alternatives to achieve its ends? (2) Have potentially discriminatory approaches to the problem been designed to maximize equity and/or compensatory justice?

Lappé *et al.* (1972) had previously recommended the following set of specific principles:

1. attainable purpose;
2. community participation;
3. equal access to testing;
4. adequate testing procedures;
5. absence of compulsion;
6. informed consent;

7. protection of subjects;
8. access to information;
9. provision of counselling/follow-up;
10. understandable relation to therapy; and
11. protection of right to privacy.

This list illustrates the considerable width of the ethical perspective.

The ethical questions relating to genetic predisposition are extremely complex. Thus, this chapter will not contain any solutions. Although ethics is sometimes looked upon as a matter of overriding moral judgment, it will be discussed as a set of considerations, where consensus needs to be obtained.

A selection of major ethical dilemmas will be dealt with:

1. equal rights in free enterprise;
2. equal right and opportunity to find work;
3. fairness towards the handicapped;
4. individual freedom;
5. rights to privacy; and
6. dealing with uncertainties.

However, to set the stage, the nature of scientific documentation will first be examined.

16.6 SCIENTIFIC DOCUMENTATION

Individual susceptibility may be classified into three groups with regard to its significance for adverse health effects of environmental exposures:

1. factors that increase the concentration of the biologically active substance at the active site;
2. factors that augment the reaction of the chemical substance with target molecules in the body, thereby initiating the response; and
3. factors that promote the sequence of events between the initial reaction and the final manifestation of an adverse health effect.

With regard to screening for such types of susceptibility, the genetic nature of the trait tested must be clearly understood, and the genetic trait must be the major factor in the development of a disease of public-health importance. Also, the potential for subsequent action must be considered; only when a screening result can lead to potential beneficial efforts should the screening test be considered (Wilson and Jungner, 1968).

The following basic requirements must therefore be assessed with regard to the screening method: scientific validity; certainty of result, and procedures for follow-up to detect false-positives and/or false-negatives; cost efficiency: degree of individual risk due to the predisposition; need for individual counselling; and proven benefit of preventive or therapeutic measures. Furthermore, any risks, for example of infection or exposure to radiation, associated with the screening method itself must also be determined.

No firm answers can be given concerning the extent to which these requirements should be fulfilled. For example, with regard to cost-efficiency considerations, Murray (1986) suggested that the relative risk for individuals with the genetic trait should be significantly increased (above 3:1) and the expected frequency of the trait should be above 1% in the population examined. Others have suggested that the expected frequency should be above 5% (Omenn, 1982). However, the basis for such general recommendations is unclear.

Obviously, the above criteria for screening methods can be difficult to satisfy in every detail. Less adherence to them would be defensible for a simple and inexpensive test that would cause no or very few ethical problems, while a less straightforward situation would necessitate a stricter documentation of test validity.

16.7 EQUAL RIGHTS IN FREE ENTERPRISE

Equal opportunity to make economic profit may result in conflicts with the responsibility towards employees and consumers. Thus, to what extent can an employer (or society) be required to create safe jobs for those individuals who turn out to be genetically susceptible? Also, to what degree should consumer products and environmental releases take into regard hypersusceptible subgroups of the general population?

An employer (or society) may obtain considerable benefits from pre-employment or preplacement genetic tests. Exclusion of susceptible workers will potentially lead to fewer job-related diseases with less sickness absence, better productivity, fewer compensation cases and smaller insurance payments. Ideally, an improved matching between job characteristics and individual genetic make-up would result in a decrease of the total and a more even distribution of risks (Lappé, 1983). The practical question is therefore to what degree the job must be fitted to suit heterogeneous human characteristics (including genetic susceptibility).

Perhaps the experience with allergenic substances can illustrate some of the difficulties. A large difference in reactivity exists between allergic individuals and the rest of the population, and the costs to attain subthreshold allergen exposures at work are often excessive. Thus, screening and employee selection would seem a cost-effective initiative. Pre-employment examination and a questionnaire regarding, for example, atopic dermatitis have therefore been used at facilities producing enzymes and other potential allergens. This practice is based on the belief that atopic individuals appear to be particularly likely to develop allergic contact dermatitis (Rystedt, 1985).

However, the economic incentive does not mean that the procedure is ethically acceptable, except perhaps from a utilitarian viewpoint. Thus, certain rights and privileges may be denied to the unfortunate individuals with a positive test, as discussed in the following section. Further, detection of genetic susceptibility in an individual can shift the focus of attention from environmental hazards to alleged or proven deficiencies of the exposed person. By 'blaming the victim', pollution abatement or other primary preventive efforts would tend to be deferred.

In a wider context, members of society other than employers may also need to take into consideration the potential susceptibility of fellow human beings towards chemical exposures. For example, exposure to tobacco smoke may constitute a risk in relation to certain forms of genetic predispositions. However, in a Swedish study, early identification of α_1-antitrypsin deficiency in newborns did not result in any desirable effect on parental smoking, although this habit is the only relevant risk behaviour in this connection (McNeil *et al.*, 1988).

16.8 EQUAL RIGHTS TO FIND WORK

Equal rights in free enterprise will invariably clash with the equal rights and opportunity to find work. The latter principle means that all individuals, regardless of sex, race or genetic make-up, should have equal opportunities in the job market. However, application of genetic screening could mean that certain privileges and rights may be denied to the unfortunate individuals with a positive test result.

Employability has hitherto been based primarily on the ability to perform a particular job, rather than the estimated long-term risks of developing a work-related disease in the future. However, these two concerns may not always be easy to separate. For example, at a pre-employment examination, the physician will only recommend males with a strong back to become butchers; and in some countries women are excluded by law from jobs handling heavy loads. This selection practice will ensure that the workers are able to carry out the heavy physical work, but it will also protect less sturdy individuals against excessive muscular strain.

Recently, some industries have become very concerned about the possible economic burdens due to compensation for occupational diseases. For example, an asbestos-manufacturing company has at some time excluded smokers from work sites with asbestos exposure because of the potentiation by tobacco smoke of asbestos-related lung cancer risk (Omenn, 1982). Also, according to international recommendations on exposure of ionizing radiation, a pre-employment examination is required to identify individuals who may be particularly vulnerable if exposed to radiation. Thus, vulnerability of the worker has already to a limited extent become an issue in hiring and placement policies. The question is whether such practices should also be allowed in relation to genetic predisposition to toxic effects caused by chemical exposures.

As with the phenotypic vulnerability, information on genetic susceptibility is likely to cause potential denial or termination of employment or transfer to another job. Thus, past experience with screening for sickle cell trait has documented that the disadvantages to those with a positive result include stigmatization, job discrimination and loss of future job opportunities (Lappé *et al.*, 1972). Furthermore, as certain genes occur more frequently in some population groups, disproportionate exclusion of, for example, members of certain ethnic and racial groups would also be a likely consequence. Not surprisingly, genetic screening evokes reactions related to ethnic distribution. Thus, although some forms of phenotypic selection violate the principle of equal right to find work, the violation potentially caused by genetic screening would be more serious.

In addition, toxic chemicals are not necessarily restricted to workplaces only. Food, consumer products and the general environment may contain chemicals to which some people have a genetic susceptibility. Thus, the right of consumers and the public at large may also need to be considered.

16.9 FAIRNESS TOWARDS THE HANDICAPPED

Based on the principle of fairness, wide efforts are made to integrate handicapped individuals in society. In some regards, these attempts are perhaps not sufficient. However, excessive efforts to accommodate individuals with particular needs or susceptibilities will clash with utilitarian approaches to achieve the greatest benefits for the greatest number.

Whether or not individuals with genetic susceptibility may be regarded as handicapped will depend on several factors. At the time of a positive test, most individuals will not exhibit any debilitating medical condition and would therefore not satisfy normal definitions of a handicap. However, these individuals are perceived to be at increased risk of developing such a condition at some future time. The individuals with positive tests may still be denied jobs and otherwise treated as if handicapped. Also, the knowledge as such may in some cases result in psychological reactions that could be regarded a mental handicap. For example, several years after identification of α_1-antitrypsin deficiency in newborns, psychological reactions were noted in both the affected children and their parents (McNeil *et al.*, 1988).

Special handicap provisions could apply, depending on the degree of physical handicap and the immediate nature of the risk. Also, the degree of discrimination and stigmatization should be taken into account, even if the genetic predisposition causes no visible impairment. If substantial, efforts may seem justified to facilitate integration of the afflicted individuals in society and prevent discrimination. The experience with α_1-antitrypsin deficiency (McNeil *et al.*, 1988) would suggest that this task is not easy.

16.10 INDIVIDUAL FREEDOM

Individual freedom to decide must be balanced against the needs for social control and paternalism. To what degree does society have a responsibility to restrain susceptible individuals for their own good and to prevent them from becoming a burden on society? The answer to this question is not simple. For example, we make it illegal to ride a motorcycle without a crash helmet, but, on the other hand, we permit smokers to continue inhaling carcinogenic tobacco fumes.

With genetic screening, public health interests would to a certain extent seem to parallel the affected individual's own interests in preserving health. Thus, the violation of individual freedom caused by the screening program could be regarded a case of less onerous, solicited paternalism. However, the scope of permissible intrusion will depend on the nature of the test and the potential for unwanted effects.

The ethical questions concerning genetic susceptibility should not be considered in isolation. Preventative medicine encompasses several approaches, including safeguarding the environment and stimulating healthy life-styles. Focusing disproportionate attention on the needs of susceptible subpopulations may distract from other efforts, with implications on preventative efforts in general. Thus, one possibility may be to identify and protect the susceptible individuals and thereby create a graded system of prevention with the associated risk of discrimination. An alternative option may be to aim at providing safe conditions for a maximal proportion of the population without considering the varying degrees of susceptibility. The choice is difficult. For example, focusing attention on susceptible groups alone could potentially penalize the remaining population, if general safeguards are deferred or cancelled.

In occupational health, some selection of phenotypes already exists, as referred to above. A higher degree of paternalism is involved when the safety of fellow employees or the general public is at stake (for example when pilots are required to be in excellent health and have good eyesight). However, perhaps apart from the requirement that policemen may not be colourblind, such selection is based on phenotypic appearance rather than genetic make-up.

The examples mentioned would appear to be in accordance with the principle that each individual should have an equal right to the most extensive liberty compatible with similar liberty for others. Also, most of the above violations of individual freedom could be considered examples of solicited paternalism that have received explicit or implicit consent from the individuals involved. However, the borderline between solicited and unsolicited paternalism is diffuse. In other words, the difficulty is to decide whether or not a governmental intervention can be considered 'weak paternalism' (Beauchamp, 1980).

16.11 RIGHTS TO PRIVACY

Openness about hazards will conflict with rights to privacy and confidentiality. To what degree do genetic tests represent bodily intrusions and invasions of privacy? Is it better to be ignorant than to be discriminated against? The right to personal privacy obviously encompasses two issues: the right to physical privacy and the right to withhold information likely to prove detrimental to one's self-interests.

Not everybody may necessarily want to know about their risks of developing a specific genetic disease. The burden of knowledge could lead to severe social and psychological consequences. For example, among individuals at 50% risk of inheriting Huntington disease, about two-thirds would plan to make use of a presymptomatic test for the gene, even when no treatment for the disease is available; on the other hand, even if treatment were available, some individuals responded that they would not want to use the test (Markel *et al.*, 1987). Similarly, a screening test for carriers of recessive gene for Tay-Sachs disease (deficient Hexosamidase B) has not become very popular among the main group at risk. Many refrain from participating, simply because a positive test result may interfere with their plans for getting married and having children.

In occupational health, eligibility for claiming compensation for job-related disease could potentially be denied those with a particular susceptibility. Then the genetic screening tests could be regarded unwanted. In such situations, one may refer to a right not to know.

In a related field, a European Community directive now requires that blood samples be drawn from workers with occupational lead exposure so that the blood lead concentration can be determined and compared to a limit value. Although venipuncture is generally regarded a procedure associated with no discernible risk, the drawing of blood will cause a small proportion to faint. More importantly, it infringes on the right to be free from an unwelcome physical intrusion and the right to make decisions regarding one's own body. Still, the benefits associated with this biological monitoring technique were considered important enough to justify its inclusion in the legislation as a mandatory test. Thus, even though protection against unreasonable search and seizure and protection of personal privacy are considered basic rights, they are obviously not inviolate.

Participation in a genetic monitoring programme could be tantamount to self-incrimination, particularly if carried out in an occupational setting. The right to preserve confidentiality may well be extended to include a right to retain the information within one's own body. Thus, major issues to be considered in this regard are whether the test is voluntary pending informed consent, and questions relating to confidentiality.

Informed consent has traditionally been used in relation to medical treatment and research, because the patient has the right to decide what is going to happen to his or her body and health. Whether or not such tests are carried out as part of a traditional physician–patient relationship, disclosure of risks and benefits would be advisable before testing is initiated. With genetic susceptibility to particular exposures, the option of making an informed decision rests on full disclosure of the significance of the genetic test result as well as full disclosure of the exposures encountered. The latter relates to the worker's right to know in occupational health. If denied that information, the benefit of an informed choice is also lost. Also, the need for more extensive counselling must be carefully ascertained.

Confidentiality is a mainstay in medical practice. An individual's right to maintain confidentiality of information regarding his or her genetic traits would place considerable limitations on the extent to which others can use this information. If the detection of genetic susceptibility results in a transfer of the patient to another job, the confidentiality is essentially non-existent. With tests for genetic susceptibility, solutions must be evaluated according to both the respect for personal privacy and the social interest in promoting public health.

16.12 DEALING WITH UNCERTAINTIES

The weaker the scientific foundation for a genetic test, the less secure is the ethical defence for opening the possibilities for violation of individual rights. Minimal difficulties in interpretation of the results should occur with tests approved for use in screening. More often, such uncertainties occur in methods used in research projects. However, although a genetic screening test has passed the formal scientific requirements outlined above, some proportion of false-negatives and false-positives will always occur. False-positives will be mistakenly labelled as genetically susceptible and will run the risk of discrimination and loss of career opportunities. False-negatives will be deprived of the benefits of added protection originally intended. When considering the possible application of a screening method, these issues must be dealt with. Retesting can to a certain extent improve the sensitivity and specificity.

In some cases, the outcome of a reliable screening test may be complex, and the risks involved may be difficult to explain. Thus, the health significance may be uncertain to the individual, and a worker with genetic predisposition may unknowingly take excess risks. In such situations, decisions on behalf of the individual could perhaps be defended. Actions must then be considered with regard to an acceptable degree of solicited paternalism, as mentioned above.

The risk involved may in some cases only become apparent at a later point in time when more information has been gathered (for example, from a continuing follow-up of exposed cohorts). Criteria and a framework for initiating comprehensive notification of workers at risk must then be prepared (Schulte, 1985, 1986).

16.13 CONCLUSIONS

In identifying cases of genetic predisposition to disease, screening tests of high diagnostic validity may potentially result in benefits due to a decreased risk of disease in employees of consumers. Tailoring of individualized protection in relation to genetic make-up would be likely to offer some benefits. Also, early diagnosis would potentially give carriers of the gene opportunity to plan their life to reduce the impact of the susceptibility. Alternatively, failure to detect and warn individuals at risk could seem culpable.

Predictive testing also promises clear benefits to those found not to be gene carriers. Although we mainly focus on susceptibility, we need to emphasize that the reverse (i.e. genetic resistance) is of equal importance.

On the other hand, testing for genetic predisposition will infringe upon individual freedom and potentially lead to violations of principles of fairness. The result of a predictive test will inevitably be of interest to others besides the person at risk, and this may lead to conflict. In this regard, the social context must be appreciated, and a balance must be sought between the interests of the individual and those of society.

The degree of risk is important. With relatively small excess risks and small proportions of the population involved, voluntary programmes would be favoured. When the risks are greater and may affect a large proportion, paternalistic intervention could be justified. However, such efforts can only be justified if a highly reliable test is available.

In a review of medical screening practices in the workplace, Ashford (1986) recommended the following strategies to deal with ethical conflicts:

1. encouragement of ethical enquiry in the conduct and use of medical screening:
2. use of legislative and regulatory means to clarify rights and duties;
3. encouragement of better disposition of conflicts by improving procedural fairness in conflict resolution; and
4. encouragement of the use of the self-help techniques (e.g choice of occupation, refusal of hazardous tasks).

These recommendations would seem to apply very well to the potential use of screening tests for genetic susceptibility. In particular, a public discourse on the ethical conflicts in this area would seem to be highly pertinent. Whether or not the goal is considered a social contract, the conflicts can only be resolved in an open, informed forum.

ACKNOWLEDGEMENTS

The author is grateful to Professor Daniel Andersen and Professor Peter Pruzan for commenting extensively on an early draft. Numerous other colleagues have offered helpful advice.

REFERENCES

Ashford, N. A. (1986) Medical screening in the workplace: legal and ethical considerations. *Semin. Occup. Med.*, **1**, 67–79.

Ashford, N. A., Spadafor, C. J. and Caldrat, C. C. (1984) Human monitoring: scientific, legal and ethical considerations. *Harvard Eviron. Law Rev.* **8**, 263–363.

Beauchamp, D. E. (1980) Public health and individual liberty. *Annu. Rev. Public Health*, **1**, 121–36.

Gewirth, A. (1986) Human rights and the workplace. *Am. J. Ind. Med.*, **9**, 31–40.

Gibbs, R. A. and Caskey, C. T. (1989) The application of recombinant DNA technology for genetic probing in epidemiology. *Annu. Rev. Public Health*, **10**, 27–48.

Lappé, M. (1983), Ethical issues in testing for differential sensitivity to occupational hazards. *J. Occup. Med.*, **25**, 797–808.

Lappé, M., Gustafson, J. and Robin, R. (1972) Ethical and social aspects of screening for genetic disease. *N. Engl. J. Med.*, **206**, 1129–32.

McNeil, T. F., Sveger, T. and Thelin, T. (1988) Psychosocial effects of screening for somatic risk; the Swedish α_1-antitrypsin experience. *Thorax*, **43**, 505–7.

Markel, D. S., Young, A. B. and Penny, J. B. (1987) At-risk persons, attitudes towards presymptomatic and prenatal testing of Huntington disease in Michigan. *Am. J. Med. Genet.*, **26**, 295–305.

Murray, R. F. (1986) Tests of so-called genetic susceptibility. *J. Occup. Med.*, **28**, 1103–7.

Omenn, G. S. (1982) Predictive identification of hypersusceptible individuals. *J. Occup. Med.*, **24**, 369–74.

Rawls, J. (1971) *A Theory of Justice*, Oxford University Press, Oxford.

Rest, K. M. and Patterson, W. B. (1986) Ethics and moral reasoning in occupational health. *Semin. Occup. Med.*, **1**, 49–57.

Rystedt, I. (1985) Work-related hand eczema in atopics. *Contact Dermatitis*, **12**, 164–71.

Schulte, P. (1985) The epidemiological basis for the notification of subjects of cohort studies. *Am. J. Epidemiol*, **121**, 351–61.

Schulte, P. (1986) Problems in notification and screening of workers at high risk of disease. *J. Occup. Med.*, **28**, 951–7.

Veatch, R. M. (1981) *A Theory of Medical Ethics*, Basic, New York.

Wilson, J. M. G. and Jungner, G. (1968) *Principles and Practice of Screening for Disease*, Public Health Papers 34, World Health Organization, Geneva.

17 Health policy, ethics and human values

Z. Bankowski

Council for International Organizations of Medical Sciences (CIOMS), Geneva, Switzerland

17.1 SUMMARY

The protection of the individual who is genetically susceptible to chemical hazards both in the workplace and in the environment raises serious ethical issues. Whether exposed populations and particularly workers should be required to undergo occupational screening for genetic susceptibility, and how the information obtained should be used, need to be approached also from an ethical point of view.

This issue is considered in the general context of health policy, ethics and human values. A health policy, whether of a whole nation or of a community, is usually formulated so as to serve both the good of the individual and the good of society. However, there is always a point at which the protection of the social good risks undermining the human values that guide health, policy-making and the rights and autonomy of the individual. Individual autonomy requires that individuals give informed consent to genetic screening and to the communication of the screening results to interested parties.

The inherent conflict can only be resolved in the context of national health policy making, informed and guided by ethicists skilled in analysing the moral content of policies and arguments.

17.2 INTRODUCTION

Many, perhaps most, health policy decisions raise ethical questions. Policies have to do with who shall receive health services, how resources shall be allocated, what criteria should be used in setting priorities, what constitutes acceptable forms of health care, when health care should be begun or ended, even the matter of who should be involved in making policy decisions, all have inherent ethical components. Different national, cultural and religious traditions yield different ethical value systems, and their interactions with health policy making will therefore vary from country to country.

Many factors other than ethics influence policy decisions. Political, economic, cultural, religious and organizational issues can obscure, distort, displace or promote ethical considerations. Of considerable importance and interest is the dynamics of policy making, whereby ethical values are taken into account or lost from sight as policies are formulated.

To provide a forum for an international and intercultural discussion of these issues, the Council for International Organizations of Medical Sciences (CIOMS) convened in Athens in 1984 an international conference on 'Health Policy, Ethics and Human Values – an International Dialogue' (Bryant and Bankowski, 1985). Participants included health policy makers, ethicists, and philosophers from different cultural and religious groups, including secularists. The conference's concerns coincided with those of the World Health Organization (WHO), particularly as regards equity, social justice, community participation and the dignity of individuals in sickness and health, as expressed in WHO's goal of health for all.

The conference helped to identify and clarify the attitudinal differences and common ethical approaches of different philosophies, cultures, traditions and religious and non-religious groups. It strongly recommended a continuing international and intercultural dialogue to improve understanding of the relationship between health policy making, ethics and human values in different cultural settings. To follow this recommendation, CIOMS undertook a new programme.

17.3 THE INTERACTION OF HEALTH POLICY MAKING, ETHICS AND HUMAN VALUES

As a basis for the discussion of the Athens conference, the American physician and philosopher, Edmund Pellegrino, (1985) explained the ways in which health policy, ethics and human values interrelate:

> The health policy of a nation or a community is its strategy for controlling and optimizing the social uses of its medical knowledge and resources.
> Human values are the guides and justifications that people use for choosing the goals, priorities and means that make up that strategy.
> Ethics acts as the bridge between health policy and values. It

examines the moral validity of the choices that must be made, and seeks to resolve conflicts between values which inevitably occur in making those choices. Ethics, therefore, orders human choices in accordance with normative principles.

A health policy, therefore, ultimately reflects the values of a human community – the fundamental beliefs and commitments that a community or nation ties most closely to its identity and integrity. Through the expression of its values in the choices and priorities of its health policies, a community exerts its influence over the enormous momentum of technological advance.

Ethics is itself grounded in sociocultural, philosophical or religious convictions. These convictions are a society's measuring rods for right and wrong, good and evil. They are often incommensurable and irreconcilable from one culture to another. To examine these convictions or to disagree with them is often perceived as a moral assault because they are so integral to a person's or a nation's self-image. It is this that makes transcultural ethical dialogue so difficult.

Health policies are rarely derived from explicit and systematic analysis of the moral values that shape them. Much of the art of national and international politics consists in structuring decision-making in such a way that value issues are not confronted. The aim is to keep peace between, and within, divergent belief systems. However, once framed, a public policy unerringly reveals the values that drive a society. This is especially the case in health policies, which so often deal with those in the 'shadows' of life – the very young, the poor, the sick, the aged or individuals who are biologically or socially vulnerable or damaged. Different peoples 'value' these groups differently. When resources are limited, 'tragic' choices must be made that may be to the disadvantage of one group or another.

In every culture there is an unavoidable ethical problem inherent in the very conception of policy formulation. This is the inevitable tension that any policy generates between individuals and social good. While most policy makers strive to serve both, there is always some point at which further improvements in a society's good intolerably undermine the worth of each individual's life within that society.

Health policies and health policy makers have three common goals:

1. To attempt to control the social and economic impact of the unrestrained use of advanced medical technology in treating individual patients.
2. To achieve a more equitable distribution of the benefits of medical knowledge.

3. To use medical knowlege in an anticipatory way for the collective good of present and future generations.

Each goal arises out of the unprecedented possibilities of modern technology, and each presents some conflict between social and individual good, between traditional medical ethics and social ethics.

The influences on those responsible for health policy are diverse – ethical, cultural, political, economic, religious, tribal, bureaucratic and personal. In any case, the policy maker will form a decision that represents his or her reconciliation of those diverse influences.

Value systems, the systems of principles or standards that determine what is valuable or important in life, play at least four different roles in health policy making.

1. They provide a framework for choice between policy alternatives. Policy makers have to consider different feasible options and decide which among them conforms best with the ethical and other values of the group.
2. They provide the framework for choosing the policy makers. To choose the policy makers is to choose the value systems upon which decisions will be made.
3. They are critical for deciding which facts will be taken into account and how they will be used for purposes of policy making.
4. They are critical in determining what the possibilities are for intercultural cooperation in health policy. There may well be intersection of these cultural traditions that facilitate – indeed require – collaboration across traditions (Veatch, 1985).

17.4 THE HEALTH PROFESSIONAL'S ROLE IN FORMULATING AND IMPLEMENTING POLICY

The formulation and implementation of health policies require the closest technical and moral cooperation of health care professionals and policy makers. Pellegrino (1985) took the physician as an example, but the same considerations apply, by and large, to all health professionals.

The physician has at least five separate levels of moral responsibility to exercise in the judgments and choices essential to policy formulation and implementation. At each level the physician interacts with policy makers, who themselves have the same range of moral obligation.

(a) The physician as the patient's advocate

The physician who offers to treat or help a person who is ill is bonded in a special way with the patient. The physician invites the patient's trust that he or she will be competent and will place that competence at the patient's disposal. Except for the most extraordinary reasons, for example the risk of serious immediate harm to others, the physician has an inescapable primary obligation to work for the patient's good, and not in his or her own or others' interests.

(b) The physician as a technical expert

When not locked into a trust relation with a particular patient, physicians are obliged to participate in policy formulation because they have the technical knowledge which policy makers need to make rational choices between alternatives. This is an obligation physicians owe society since their knowledge is not proprietary. A clear separation must be drawn between the physician's role as expert in medical science and his or her presumption of expertise in value decisions. In this respect physicians face the same problem all experts face: how to separate what is known factually and objectively from what they would like to see enacted as policy.

(c) The physician as citizen

The physician's duty to act as a technical witness in policy formulation does not exempt him or her from personal obligations as a citizen. Thus, physicians must favour policies that are morally sound, and oppose those that are immoral. As long as physicians do not take advantage of their position as technical authority they can, and should, take the measures open to all citizens in society to work for those health policies that will benefit their patients or society at large. In this the physician has the same moral obligation as any other citizen and with no more or less authority.

(d) The physician implementing health policies

Sooner or later, health policies must be implemented in the care of individual patients. Here the physician faces the most difficult moral conflicts and the most direct challenge to the traditional medical ethic of beneficence. Several positions have been taken with respect to this moral dilemma: the physician as primarily an agent of social or fiscal policies; the physician as arbiter of the application of health policies; the physician as protector of the patient's autonomy. Whichever position one chooses, the central moral issue remains: when there is conflict of moral values in implementation of a health policy, whose good should obtain – the patient's, the physician's or society's as expressed in a health policy? This is a challenge to all physicians, and in all cultures, since in all cultures there is still a strong remnant of primary duty and dedication to the interests of the one who is ill.

(e) The physician as moral agent

The physician, like any other person, is accountable. Therefore, the physician cannot carry out a health policy – for example, mandatory sterilization of carriers of genetic disorders, or refusal of care because of inability to pay – that he or she considers fundamentally immoral. The physician has a moral responsibility that transcends his role as an agent of public policy.

The traditional ethical responsiblities of the physician towards the individual patient apply also today in the professional relationship of physicians towards groups of people who cannot be called patients but whose health is to some extent the physician's responsibility. A case in point is that of the occupational health physicians, who have a professional relationship with industrial workers and a responsibility towards them to protect their health from hazards in the work environment. Their competence is placed at the workers disposal and for their good. They have obligations as citizens to favour morally sound, and oppose morally unsound, policies. When there is a conflict of moral values in application of a health policy they must as physicians take a position, and as moral agents cannot carry out what they consider an immoral health policy.

Health policy makers have the same five moral roles in policy formation and implementation. They are the agents of individual and social good. They too are technical experts. They too are morally responsible citizens, agents of policy implementation and morally accountable persons. Policy makers have not traditionally had professional codes to guide their ethical decisions. Perhaps the time has come for them too to recognize explicitly the obligations they owe their fellow humans for the moral quality of their actions and decisions.

17.5 ETHICAL ASPECTS OF GENETIC SCREENING

Research at present under way into molecular genetics, and particularly into human genome mapping and sequencing, presages a new scientific era in the medicine of the 21st century. However, there is much concern about the implications of advances in human genome mapping, genetic screening and genetic therapy for the rights and responsibilities of individuals and societies.

Efforts have begun in the USA, Japan and Europe to map and sequence the entire human genome. Over the next decade most of the genes of the human genome are likely to be identified. This knowledge will be important not only for diagnosing, treating and even preventing diseases caused by single-gene defects, but also for better understanding, and even treatment, of those major common diseases that result from the interraction of genetic predisposition with various environmental factors.

These scientific developments will raise difficult ethical issues. Increasingly, single-gene disorders will be diagnosable prenatally as well as postnatally. However, there are questions about the extent to which such screening should be implemented. At present, very few single-gene disorders can be treated. Even if gene therapy becomes feasible, it is very likely for many years to be limited to a few diseases. Hence, results of prenatal and postnatal screening will be used primarily for genetic counselling for decisions regarding the abortion of affected fetuses.

Ethical issues are equally important in the case of multifactorial diseases. The possibility of detecting individual predisposition to a given disease must be weighed against the risk of misuse of this information.

Another ethical issue arises when genetic diagnostic advances come faster than advances in treatment. What is the use to an individual of knowing of a predisposition to a disease if there is no way of preventing or treating the disease? Can genetic screening be justified by the possibility of postponing onset of disease or lessening its effects?

Two areas of concern in this context are the ethical and policy aspects of genetic screening and counselling and of DNA technologies, and the social risks associated with occupational screening for genetic susceptibility to hazards in the workplace. Two examples may be considered of the possibilities and dilemmas raised by genetic screening and counselling: one relating to the ownership of genetic information and the other to the uses of genetic screening and engineering and the caution demanded by their uses.

Traditionally the moral argument with regard to the ownership of genetic information has been that the individual has an absolute right to professional secrecy and the right to know the results of genetic screening. There is also an acknowledged right not to know, for one reason because of the possibility of misuse of information obtained from predictive testing. However, a new issue today, in the light of advances in predictive technology, is whether one's right to receive information also contains a duty to receive information about the future prospects of one's life, since what one does with one's life inevitably affects the lives of others. It seems to be a fundamental principle of the individual's autonomy that there is a basic right to be or not to be tested and to choose to be informed of the result fully or only in part. The ownership of genetic information is an area of interaction between advances in biomedical sciences, ethics and health policy making which needs further elaboration. It is increasingly accepted that predictive screening is warranted only if the individuals concerned are in a position to make a free and informed decision whether to be tested, and if adequate facilities and experiences are available for long-term follow-up of those identified at an early stage as carriers of disease genes or as individuals with genetic predisposition to disease.

The use of genetic tests might become required – or be proposed – for applications for jobs or insurance cover. This would extend the present problem that many prognostic risk data are available through a simple family history, whereas many countries have laws enabling both future employers and insurance companies, especially for life and disability insurance, to ask certain questions, and obliging the applicant for employment or insurance cover to answer. Individuals carrying a certain genetic risk might perceive a conflict of beneficence towards themselves or their families and towards their employer and co-workers. Exclusion from work, for example, because of risk of genetic disease, is warranted only if the onset of the disease would induce a severe risk to the company or to other workers, if a high-accuracy test is available and if only a limited number of people would be excluded (Niermeijer, 1988).

The converse position is the protection of the genetically predisposed individual from chemical hazards in the workplace. Should applicants be required for their own good to submit to genetic screening? Should those who test positive accept the responsibility to decide for themselves whether or not to accept the employment? What are the obligations of the occupational health physician *vis-à-vis* the employer or the applicant's personal doctor? What precautions must be taken to preserve complete confidentiality with regard to such an intimate matter?

The problems raised by advances in genetic technology need discussion in individual societies between politicians, government, employers, employees and patient organizations, with support from geneticists, ethicists and lawyers. Otherwise, many more people might become socially handicapped because of inability to obtain employment or insurance necessary for social functioning, only because of certain genetic predisposition. Those who would be identified, and only by chance, would be carrying the 'burden' of genetic disease, whereas everybody carries some five to ten deleterious genes. The responsibility of the human genome must be shared in a more balanced way. This will also preserve the acceptability of genetic diagnostic and predictive facilities, which have created new options for individual choices in life and in reproduction, beneficial for many individuals and couples at risk.

Today it is possible and indeed necessary to examine ethical questions not only as ethics of care, but much more extensively as the ethics of health policy and biotechnology as a whole. The intention is not to undervalue moral and ethical aspects at the individual level, but rather to view them within a wider frame of reference.

REFERENCES

Bryant, J. H. and Bankowski, Z. (eds) (1985) *Health Policy, Ethics and Human Values – An International Dialogue. Highlights of the Athens Conference*, Council for International Organizations of Medical Sciences, Geneva.

Niermeijer, M. F. (1988) Genetic screening and counselling – implications of the DNA technologies, in *Health policy, ethics and human values: European and North American perspectives* (eds J. H. Bryant and Z. Bankowski), Council for International Organizations of Medical Sciences, Geneva.

Pellegrino, E. (1985) Keynote address: Health policy, ethics and human values, *Health Policy, Ethics and Human Values – An International Dialogue* (eds J. H. Bryant and Z. Bankowski), Council for International Organizations of Medical Sciences, Geneva.

Veatch, R. (1985) Value systems: their roles in shaping decisions, in *Health Policy, Ethics, and Human Values – An International Dialogue* (eds. J. H. Bryant and Z. Bankowski), Council for International Organizations of Medical Sciences, Geneva.

18 Significance of hypersusceptibility for risk assessment

C. Sonich-Mullin

Environmental Criteria and Assessment Office, US Environmental Protection Agency, Cincinnati, OH, USA

18.1 SUMMARY

Consideration of hypersusceptible subgroups in the risk assessment process has traditionally been a qualitative effort. Little consideration has been given to quantitatively or methodically incorporating the risk to specific hypersusceptible subgroups in chemical-specific risk assessments. This chapter assesses the need to identify and protect such individuals by looking at exposure to toxic waste sites. The extent to which the present approach protects hypersusceptible subgroups is discussed and the proportion of the population comprising hypersusceptible individuals is determined.

Given the large number of individuals identified to be at high risk, the present approach may only be adequate for the assessment of risk associated with exposure to single chemicals. An alternative approach is necessary when evaluating health risks associated with exposure to multiple chemicals – the realistic exposure scenario. As the importance of considering such groups is evident, a scheme has been devised to assess sytematically the vulnerability of high-risk groups to exposure to multiple chemicals for a defined site. This overall scheme is envisaged as a component of the multiple chemical risk assessment procedure.

18.2 INTRODUCTION

The existence of hypersusceptible individuals has long been recognized by the United States Environmental Protection Agency (EPA), even in the absence of specific data on the response of hypersusceptible humans or experimental animals. However, when specific subgroups have been identified, they are often only considered qualitatively in criteria derivation. Little consideration has been given toward the methodical protection of specific hypersusceptible subgroups. This chapter critically assesses the need to identify and protect such individuals, particularly in the context of risk assessment following exposure to toxic waste sites. In this regard, two main issues will be addressed:

1. To what extent does the present approach protect high-risk, including hypersusceptible, subgroups of the population? (section 18.3)
2. Do these hypersusceptible individuals comprise a proportion of the population that is sufficiently large to justify the systematic consideration of high-risk groups in the risk assessment process? (section 18.4).

Prior to addressing these issues, specific terminology and background must be provided. Hypersusceptibility and sensitivity have been used interchangeably. Redmond (1981) defines sensitivity as 'responsiveness to a pollutant', where sensitivity refers to the rate of change of a response as the dose increases. We prefer the use of hypersusceptibility in that it simply implies 'more susceptible'. A hypersusceptible individual is one who will experience an adverse health effect, to one or more pollutants, significantly before the general population because of one or more factors present that predispose that individual to the harmful effects.

These individuals are essentially at higher risk of adverse health effects due to exposure to the pollutants. However, 'high risk' is often used to designate specific groups, such as occupational groups, that are at higher risk because they are exposed to higher exposures. Another example would be persons living near hazardous waste sites. The distinction here is that some of the workers or some of the residents may be hypersusceptible but all workers and all residents are at high risk because they have been exposed to levels higher than those to which the general population is exposed.

In *Pollutants and High-Risk Groups*, Calabrese (1978) classifies hypersusceptible individuals into five main categories based on biological factors that increase human susceptibility to pollutants, as described in Table 18.1. The first four categories will mainly be focused upon in this chapter. Known physiological mechanisms form the basis of Calabrese's categorizations and include phenomena such as transplacental transfer of certain chemicals, greater gastrointestinal absorption in younger children and vulnerability of the embryo and fetus. These factors clearly indicate that there is a wide range of susceptibility in the human population.

Table 18.1 Biological factors predisposing individuals to hypersusceptibility to pollutants (adapted from Calabrese, 1978)

Factor			*Rationale*
I	Developmental processes		
	Examples:	Embryos, fetuses and neonates	Immature enzyme detoxification systems
		Young children	Higher rate of gastrointestinal absorption
II	Genetic disorders		
	Examples:	G–6-PD deficiency	Haemolysis in presence of certain chemicals
		Sickle-cell trait	Predisposes toward haemolytic anemia
III	Nutritional deficiencies		
	Examples:	Vitamin C deficiency	Potentiates effects of pollutants
		Protein deficiency	Affects metabolism of insecticides
IV	Existing disease		
	Examples:	Heart	Increased mortality in pollution episodes demonstrated in epidemiological studies
		Lung	Existing conditions are aggravated by respiratory irritants
V	Behavioral factors		
	Examples:	Smoking	Increases exposures and therefore risk; interferes with normal cleansing mechanisms of lung
		Alcohol	Liver damage, synergism with other chemicals
		Dietary patterns	Increases exposures; deficiency states increase risk

Few data exist to help quantify the amount of excess risk due to hypersusceptibility in humans. Such quantification can usually be inferred from appropriate animal data; however, such data are scarce. The qualitative data as well as the mechanistic processs support the existence of such groups. Qualitative support is derived from observations in both animals and humans that adverse effects occur in only a proportion of those exposed, in spite of similar exposures. The effects observed in the exposed population may also vary in degree of severity.

Despite the lack of quantitative dose-response information, hypersusceptible individuals have not been totally neglected in regulatory procedures. For example, the Clean Air Act of 1970 mandated that the health of hypersusceptible as well as healthy individuals of the population be protected. Two instances for which standards have been set for the hypersusceptible individual are nitrates and lead. For both contaminants the hypersusceptibles are infants and children. In the case of nitrates in particular, infants are the only group of the population affected. It appears that standards are set for the hypersusceptible group only if available human data demonstrate the risk to that particular group. If, however, animal models and/or mechanistic information suggest a hypersusceptible group, then this information can only be considered qualitatively in scientific guidance for risk assessment (for example, the effect of PCBs (polychlorinated biphenyls) or chlorinated hydrocarbons on pregnant women).

In general, the neglect of such subgroups is due to the lack of quantitative data and the assumption that those individuals comprise only a small proportion of the population. This latter issue will be discussed and assessed in detail in section 18.4. However, although some of these groups may be quite small in number for a given chemical, the issue must be considered in light of potential exposure to complex mixtures. Since one or more specific hypersusceptible subgroups may be associated with each contaminant, the total number of individuals classified as hypersusceptible could comprise a significant proportion of the population. Thus, multiple chemical exposures could lead to multiple hypersusceptible subgroups. Furthermore, because of the diversity of each hypersusceptible subgroup, it has been postulated that everyone in the population will be a member of a susceptible group at certain times during life (Calabrese, 1978; Redmond, 1981).

18.3 DOES THE PRESENT APPROACH PROTECT HYPERSUSCEPTIBLE SUBGROUPS?

The US EPA guidelines suggest that a 100-fold uncertainty factor be incorporated when extrapolating from chronic animal studies to humans. This can be perceived as including a 10-fold factor to account for interspecies variability (i.e., extrapolation from animals to humans) and a 10-fold factor to account for intraspecies variability (i.e., the average versus the sensitive human). To evaluate the extent to which the present approach protects hypersusceptible subgroups, the data source as well as the extrapolation process must be considered.

For the majority of chemicals, the sources of the data are laboratory experiments. Experimental animals are in-bred for physiological homogeneity. Further homogeneity is introduced by experimental designs that select animals of similar age and weight.

In an attempt to ascertain the extent to which a 10-fold dose reduction protects sensitive members of the animal population, Dourson and Stara

(1983) examined the range of log-dose probit slopes of 490 animal studies compiled by Weil (1972). The majority of the slopes, based on acute effects, were greater than 3. This would require a dose reduction of 10 to drop the average response at least three standard deviations, that is, to a level protective of the sensitive animal (Weil, 1972; Dourson and Stara, 1983). For those 8% of chemicals in the example having probit log-dose slope less than 3, a 10-fold reduction in dose would not achieve a three standard deviation reduction in response. Thus, the 10-fold factor may protect the sensitive laboratory animal, based on an assumed normal distribution of variability for a majority of chemicals. There are few similar studies of the dose-response relationship in humans.

This 10-fold factor may not be adequate to protect humans because the human population is not as homogeneous as the animal population. Contributing to the heterogeneity in human populations is the fact that all members of the population are exposed to environmental pollutants regardless of health status, sex, age, weight or nutritional status. The range of individual variability of physiological values is ill-defined and is often not normally distributed in the healthy population. This variability, for those who are diseased, is generally much larger than that of the healthy population. The uncertainty factor concept was developed in recognition of this variability as well as other uncertainties of the extrapolation procedure. However, the selection of the number 10 is not based on a quantitative assessment of this variability.

The present approach has not been designed in ignorance of the existence of hypersusceptible individuals but rather on the assumptions that (1) no quantitative data are available to support a more accurate dose reduction to protect them, and (2) that, for any one chemical, the proportion of the population that is hypersusceptible is too small to merit consideration. The issues surrounding the former assumption warrant further study. The issues surrounding the latter assumption are evaluated in the following section.

18.4 WHAT PROPORTION OF THE EXPOSED POPULATION IS HYPERSUSCEPTIBLE?

The hypothesis is that for a population exposed to multiple chemicals the proportion that can be considered hypersusceptible will be substantial. Various subgroups will be hypersusceptible to at least one of the chemicals and some to more than one. Therefore, while the present approach may be adequate for single chemicals, an alternative is necessary for the case of exposure to multiple chemicals.

There are two components to the process of quantitatively estimating the number hypersusceptible to toxic effects from exposure to multiple chemicals. The first step is the qualitative identification of those hypersusceptible subgroups. The second is the estimation of the frequency of these groups.

18.4.1 QUALITATIVE IDENTIFICATION OF HYPERSUSCEPTIBLE GROUPS

The hypersusceptible subgroups to be considered in this chapter have been identified earlier in the chapter and they will be expanded upon here. The mechanistic approach of Calabrese (1978) is particularly useful due to the absence of human and animal data to illustrate hypersusceptibility empirically for specific chemicals. Calabrese (1983) recognizes the scarcity of laboratory data and suggests potential animal models, several diseases and conditions of proposed hypersusceptibility. Cardiovascular disease is the leading cause of death in the USA and a highly prevalent cause of morbidity. Although this disease has been widely studied, McCauley and Bull (1980) contend that the impact of environmental contamination on cardiovascular toxicity is not well known. Toxicological and epidemiological studies have shown an association between cardiovascular disease and environmental contaminants such as certain metal ions and chlorinated solvents. In some of the animal studies healthy animals are used and exposure levels are quite high. On the other hand, human studies are unable to control for confounding factors such as diet and smoking habits. McCauley and Bull (1980) suggest animal models for various cardiovascular pathologies. Thus, animal studies offer the opportunity to control for the complexity of factors in the human environment and to define specific steps in the disease process that environmental toxins exacerbate. These authors further suggest animal models appropriate to the study of such environmental contamination.

The category of hypersusceptibility that has received the most attention in cases of environmental pollution is teratogenicity. The biological basis for embryonic and fetal vulnerability and for transplacental carcinogenicity is strong (Strobino *et al.*, 1978; Fabricant and Legator, 1981; Kurzel and Cetrulo, 1981; Rao *et al.*, 1981; Rice, 1981; Saxena *et al.*, 1981). However, the relationship between animal and human teratogenicity is not as clearly established as the relationship between animal and human carcinogenicity.

Through preliminary investigations we have identified several chemicals to support the hypothesis that several subgroups are hypersusceptible to environmental exposures. Table 18.2 provides a list of these chemicals. In addition to the embryo, fetus and young children, those with existing pulmonary disease and coronary heart disease are likely candidates for high prevalence subgroups.

Table 18.2 Prevalence of subgroups hypersusceptible to effects of common pollutants

Hypersusceptible	*Prevalence*[a]	Chemicals[b]	Reference[b]
Embryo, fetus, neonate	Pregnant women: 21/1000	Carcinogens, solvents, CO, mercury, lead, PCBs, pesticides	Kurzel and Cetrulo, 1981; Rice, 1981; Saxena *et al.*, 1981
Young children	Ages 1–4: 70/1000	Hepatotoxins, PCBs, metals	Calabrese, 1983
Lung disease	Emphysema, asthma: 37/1000	Ozone, Cd, particulates, SO_2, NO_2	Redmond, 1981
Coronary heart disease	Coronary heart disease: 16–27/1000	Chlorinated solvents, fluorocarbons	Aviado, 1978; McCauley and Bull, 1980
Liver disease	Liver condition: 20/1000	Carbon tetrachloride, PCBs, insecticides, carcinogens	Calabrese, 1978

[a]All estimates based on 1970 US Census data (National Centre for Health Statistics, 1974 and 1975)
[b]Representative sample; some evidence for animal studies only.

18.4.2 Estimates of prevalence of hypersusceptible subgroups

While statistics on births and mortality are routinely collected, statistics on morbidity are not. The prevalence of chronic conditions is available from epidemiological studies and national surveys (for example, the Framingham study and the National Health and Nutrition Examination Survey), which are not collected by political division as are vital statistics. Furthermore, the epidemiological surveys are irregularly performed on a national sample, whereas vital statistics are compiled every year for the entire population.

Table 18.2 also provides the prevalence of several specific hypersusceptible subgroups. Morbidity rates from an interview survey conducted by the National Center for Health Statistics (1974) were chronic bronchitis, 33 per 1000; hypertensive disease, 60 per 1000; and heart conditions, 50 per 1000. This interview did not include residential institutions such as nursing homes and penitentiaries and is believed to underestimate the true prevalence in the general population. Therefore, the potential number at risk may be higher.

The proportion of hypersusceptible individuals in an exposed population may be estimated by multiplying the exposed population by the proportion at risk in the general study population. Although greater precision could be obtained by partitioning the exposed population into demographic subgroups at differing risks and calculating age-, race- and sex-specific risks, this is generally believed to be impractical. Thus, estimates of the prevalence of hypersusceptible subgroups for a given site will be imprecise. For this reason, a risk assessment scheme for cases of exposure to multiple chemicals may be limited to an ordinal or ranking scheme, rather than one which is strictly quantitative.

When exposure to more than one chemical is considered, the number is even greater. Table 18.3 provides inventories of typical waste sites along with the prevalance of the high-risk subgroups. One method of estimating the proportion hypersusceptible to more than one chemical is to add the rates for each chemical. The proportion at high risk is conservatively estimated to be from 10 to 20% among these sites. The number of higher-risk individuals for a hypothetical population of 5000 is given in the last column. Clearly, some of the categories of chronic disease are not mutually exclusive and addition would provide an overestimate since some individuals can be the victims of more than one condition. On the other hand, certain individuals may be excessively hypersusceptible to the effects of one chemical due to more than one pre-existing condition.

CONCLUSIONS

Due to the large number of individuals in high-risk subgroups the present approach may be adequate only for single chemicals. An alternative approach is necessary to evaluate the health risk from exposure to multiple chemicals. Some of the hypersusceptible subgroups for single chemicals may comprise a significant proportion of the population, while others may not.

However, when multiple chemical exposures are considered and, therefore, multiple hypersusceptible subgroups, even the very small groups become significant because they are part of a large number of individuals who are hypersusceptible to the chemical mixture. For a typical waste site, 10–20% of the population may be at high risk.

Because the importance of considering such groups is evident, a scheme has been devised to assess systematically the vulnerability of high-risk groups to exposure to multiple chemicals for a defined site. The scheme is proposed in order to derive an index for a hazardous waste site on an ordinal scale. The proposed approach involves the following steps.

1. Obtain monitoring data relative to the exposure.
2. List the hypersusceptible subgroups associated with each contaminant.
3. List the prevalence rates for each hypersusceptible subgroup.
4. Calculate (from 2 and 3) the estimated percentage and number of hypersusceptible individuals in the exposed population.
5. Incorporate this index into the site-specific risk assessment.

Thus the scheme commences with identification of the contaminants and their respective hypersusceptible subgroups. Once these subgroups have been identified, the proportion of the population they comprise will be estimated by applying prevalence rates for the general population to the exposed population. Methods of merging the affected population with the exposure data will be explored. This overall ranking scheme is envisaged as a component of the multiple chemical risk assessment procedure.

DISCLAIMER

The views in this chapter are those of the author and do not necessarily reflect the views or policies of the US Environmental Protection Agency.

Table 18.3 Hypersusceptible subgroups associated with typical inventory of chemicals at waste sites to which people have been exposed

Occurrence	*Chemical*	*Hypersusceptible subgroup*	*Prevalence rate (per 1000)*	*No. of hypersusceptibles in hypothetical population of 5000*
4 out of 4 sites	Chlorinated ethanes (dichloroethanes)	CHD[a]	CHD: 24	120
		Liver condition	Liver condition: 20	100
		Pre-exposure to hepatotoxins	Lung (bronchitis): 33	165
		Embryo/fetus	Pregnant women: 21	105
	Dichloroethylenes	CHD		
	1,1,1-Trichloroethane	CHD		
		Liver condition		
		Pre-exposure to hepatotoxins		
	Trichloroethylene	Liver condition		
	Chloroform	CHD		
		Lung		
3 out of 4 sites	Benzene	Thalassemia	Thalassemia: 0.1–8% of persons of Italian, Greek, Syrian and African origin	
		CHD	CHD: 24	120
		Others		
	Ethyl benzene	Thalassemia		
		Liver	Liver condition: 20	100
		Lung	Lung (bronchitis): 33	165
		Others		
	Methylene chloride	CHD		
		Liver condition		
	Tetrachloroethylene	Lung		
		Liver condition		

2 out of 4 sites	Carbon tetrachloride	CHD	CHD: 24	120
		Liver condition	Liver conditon: 20	100
		Lung	Lung (bronchitis): 33	165
	Trichlorofluoromethane	Lung		
	Vinyl chloride	CHD		
	Arsenic	Vitamin C deficiency	Vitamin C deficiency: 10–30% of infants, children and adults of low income	25
	Chromium	Young children (to metals)	Ages 1–4: 70	350
	Selenium	Vitamin E deficiency		
	Silver	Selenium defiency		
	Mercury	Vitamin C deficiency, cystinuria, pregnant women plus others		105

[a]CHD: chronic heart disease.

ACKNOWLEDGEMENT

The author wishes to acknowledge the efforts of Dr Linda S. Erdreich in developing this manuscript.

REFERENCES

Aviado, D. M. (1978). Effects of fluorocarbons, chlorinated solvents, and inosine on the cardiopulmonary system. *Environ. Health Perspect.* **26**, 207–15.

Calabrese, E. J. (1978) *Pollutants and High-Risk Groups. The Biological Basis of Increased Human Susceptibility to Environmental and Occupational Pollutants*, Wiley, New York.

Calabrese, E. J. (1983) *Principles of Animal Extrapolation*, Wiley, New York.

Dourson, M. L. and Stara, J. F. (1981) Regulatory and experimental support of uncertainty (safety) factors. *Regul. Toxicol. Pharmacol.* **3**, 224–38.

Fabricant, J. D. and Legator, M. S. (1981) Etiology, role and detection of chromosomal aberrations in man. *J. Occup. Med.* **23**, 617–25.

Kurzel, R. B. and Cetrulo, C. L. (1981) The effect of environmental pollutants on human reproduction, including birth defects. *Environ. Sci. Technol.*,**15**, 626–40.

McCauley, P. T. and Bull, R. J. (1980) Experimental approaches to evaluating the role of environmental factors in the development of cardiovascular disease. *J. Environ. Pathol. Toxicol.*, **4**, 27–50.

National Center for Health Statistics (1974) *Prevalence of Chronic Circulatory Conditions, United States, 1972.* Vital Health Statistics Series 10, No. 93, Public Health Service, US Government Printing Office, Washington.

National Center for Health Statistics (1975) *Selected vital Health Statistics in Poverty and Nonpoverty Areas of 19 Large Cities, United States, 1969–71.* Vital Health Statistics Series 21, No. 26, Public Health Service, US Government Printing Office, Washington.

Rao, K. S., Schievetz, B. A. and Park, C. N. (1981) Reproductive toxicity risk assessment of chemicals. *Vet. Hum. Toxicol.*, **23**, 167–75.

Redmond, C. K. (1981) Sensitive population subsets in relation to effects of low doses. *Environ. Health Perspect.*, **42**, 137–40.

Rice, J. M. (1981) Prenatal susceptibility to carcinogenesis by xenobiotic substances including vinyl chloride. *Environ. Health Perspect*, **41**, 179–88.

Saxena, M. C., Siddiqui, J. K. J., Bhargaua, A. K. *et al.* (1981) Placental transfer of pesticides in humans. *Arch. Toxicol.*, **48**, 127–34.

Strobino, B. R., Kline, J. and Stein, Z. (1978) Chemical and physical exposures of parents: effects on human reproduction and offspring. *Early Hum. Dev.*, **1**, 371–99.

Weil, C. S. (1972) Statistics vs. safety factors and scientific judgement in the evaluation of safety for man. *Toxicol. Appl. Pharmacol.*, **21**, 454–63.

19 Examples of regulatory practices to protect hypersusceptible groups of workers

F. Kaloyanova

Department of Toxicology, Institute of Hygiene and Occupational Health, Sofia, Bulgaria

19.1 SUMMARY

Experimental and, to a lesser extent, epidemiological evidence has demonstrated age- and sex-related differences in individual susceptibility to toxic effects of chemicals.

To protect hypersusceptible individuals, legislation has been enacted in several countries; the main target groups are women and children. Ideally, exposure limits and other preventive measures should aim at protecting the workers who are most sensitive to toxic effects of chemicals. However, satisfactory protection may not always be achieved. Thus, for example, in some Eastern European countries, women and adolescents are not allowed to work in particular industries or in jobs with exposures to specific chemicals that are considered harmful (for example to pregnant women). None of the legislation reviewed included any restrictions with regard to genotype.

19.2 INTRODUCTION

The susceptibility of the organism to chemicals depends on several factors that are incompletely known. However, many studies indicate that women are more susceptible to certain chemical exposures than men.

Many experimental, and to a lesser extent epidemiological, data show sex-related susceptibility to toxic substances. The significance of gender is reflected first in the effects of oestrogenous and androgenous hormones on the metabolizing enzymes. The microsomal metabolizing capacity of liver is higher in male rats for a number of toxicants. Male sex hormones stimulate microsomal enzymes. The agent that is activated in vivo by the same enzyme systems may be less toxic in female rats.

Sex differences in penetration, excretion and accumulation of some substances have also been found. Because of some sex differences in the morphological composition of blood, a more frequent breakdown in the adaptation processes occurs in women, with the development of anaemia and other changes in the health state (Beliles, 1972; Tolokontsev and Filo, 1976; Nikiforov and Tabakova, 1981).

The higher fat tissue content in the female organism determines a greater deposition of toxic substances. Females are more resistant to chemical agents that produce circulatory stress (WHO, 1987).

Krassovski (1973) performed experiments with 149 substances upon Wistar rats and established 2- to 3-fold differences between males and females, with the exception of organophosphorous compounds, to which females proved to be four times more susceptible than males. Other studies on various species of experimental animal have also shown gender differences. A typical example of the higher sensitivity of females is the myelotoxic action of benzene. Females are more susceptible to parathion, certain organic solvents and boric acid, as well as to substances causing specific and general effects in blood, such as lead, cadmium, arsenic, styrene and chloroprene. Females are less susceptible to ethyl alcohol, mercury, chloroform, aflatoxin B and acetylaminfluorene (de Bruin, 1976; Panova, 1988). Calabrese (1985) has performed thorough analyses of sex differences in terms of toxic susceptibility. In the majority of cases, the data are obtained in experiments with laboratory animals. A higher susceptibility of female animals to lead (resulting in impairment in haemoglobin formation, in particular), DDT (more efficient storage), organophosphate insecticides and hexachlorobenzene was found.

Male animals are more sensitive towards ammonia, cadmium, copper and selenium.

19.3 REGULATORY PRACTICES

To protect susceptible workers against toxic effects, legislation has been enacted in different countries. In particular, women and children are covered by such laws. Age may also be considered as a factor for potentiation of some genetic predisposition. Zielhuis and Wibowo (1989) point out that health-based recommended occupational exposure limits may vary according to gender. The occupational exposure limit should be adjusted to the sex that is most at risk.

19.3.1 Regulations to protect young people

Raxman and Zuckermann (1987) point out that laws concerning employment of children have been enacted for three fundamental health-related reasons, namely to prevent children from being engaged in perilous and hazardous occupations, to protect them from injuring themselves because of their inexperience and carelessness and to prevent overworking during an important phase of their physical and mental development.

In the Code of Labour in force in Bulgaria, limitations are specified for young people and women working with chemical substances that could represent a health hazard for them. No special regulation refers to phenotype. The minimum age of employment is, as a rule, 16 years. According to a list, supplementary to Regulation No. 4, published in *State Gazette* No. 44 in 1987, individuals under the age of 18 years are excluded from 134 jobs involving exposure to toxic substances. The latter include lead, mercury and arsenic compounds, sulfurous and nitric acids, calcium hypochlorite and chlorine, nitrocellulose lacquers, pesticides and mineral fertilizers, synthetic fibres and plastics. The list also includes a number of processes in the oil and chemical industries, textile and printing industries, mining, metallurgy, glass, woodworking and furniture industries, food processing industry and transport industry.

Poland has similar documents governing the employment of adolescents (Rozpordzazenie Minstra Pracy, 1979).

In the USSR, several thousand types of industrial work are categorized as dangerous for persons under the age of 18 years (USSR State Committee for Labour and Social Affairs, 1989).

In Czechoslovakia the restrictions for youth are given in separate legislation for the different branches of industry (Osobotná úprava pracovnysch podmienik zien, 1972). A great number of chemical substances are included.

Legislation varies between countries and is consequently preventing exposure of youth to noxious chemicals difficult to compare. The common factor is that the protection covers youth up to the age of 18 years.

19.3.2 Regulations to protect women

In many countries, women of childbearing age have a greater protection than men from contact with toxic substances.

The Code of Labour in Bulgaria includes clauses for the protection of women. For example, performance of heavy work or work that is harmful to a mother's health and maternal functions is forbidden. Such categories of work are defined in a special list. The first sections include the kinds of production from which women are excluded, independent of their age. These include mining and processing of some metals (for example lead and cadmium), working with carbide, benzene, nitro- and aminoderivatives of benzene, production of trinitrotoluene and mercurial explosives and the casting of lead type.

The second part of the list includes types of production from which women who intend to bear children are prohibited. These women are identified by their personal declaration. These types of production involve exposure to manganese and its compounds, arsenic salts, lead and chromium dyes and pigments, aniline dyes (production), hydrofluorine vapours, pesticides, vinylchloride and carbon disulfide. Work with selenium and its compounds, dimethylformamide and ethylene oxide (if their concentrations exceed the MAC (Maximum allowable concentrations) values or are carcinogenic or mutagenic) is also prohibited.

In Poland (Wykaz prac wzbronionych kobietom, 1979; Rozporzadzenie Rady Ministrow, 1984) regulations for the protection of women also exist. Three groups of women are specified (up to 35, 36–45 years and pregnant women) and the concentration of the toxic chemical in the workplace is controlled. For example, women are not allowed to work with mercury compounds, benzidine, 1-naphthylamine nor aniline dyes. In addition, when the concentrations exceed the MAC values, work with benzene, chlorinated aliphatic hydrocarbons such as chloroform, 3-chloroethylene, 4-chlorethylene and 4-chlormethane, acetone, carbon disulfide, alcohol, formaldehyde, organic dyes and nitrogen oxides is prohibited.

In Czechoslovakia, regulations enumerate in detail the industrial branches from which women are prohibited. Contact with a great number of chemicals is consequently avoided.

In the former German Democratic Republic, people up to 16 years of age were not allowed to work where concentrations of gases or vapours approached the MAC values, in production or processing of sex hormones or when a risk of infectious disease was present. Under discussion is legislation that would contain the following protective clauses:

- to prohibit young people and women from working where concentrations exceed the MAC values;
- to allow young people under 16 years of age to work only where concentrations do not exceed the 0.5 MAC value;
- to prohibit young people from working with mutagens, carcinogens and teratogens;
- to allow work with chemical substances of the first class (highly toxic) only after 16 years of age and during training.

In Hungary, a decree of the Ministry of Health forbids young people and pregnant women to work with toxic pesticides, and forbids women and young people to work with substances that exceed their MAC values of $\geqslant 5\ mg/m^3$.

Young people and pregnant women are prohibited from working with some substances even when exposure is minimal. Such substances include arsenic compounds, anticoagulants, benzene, asbestos, benzidine, cytostatics (production), phenol, mercury and its compounds, nicotine, lead and its compounds and carbon sulfide. When women are hired they are instructed to inform the factory physician when they plan a pregnancy (cease the use of contraceptives), so they may be removed from exposure to chemicals.

19.4 CONCLUSIONS

A comparative analysis of the legislation in the countries cited above reveals that, in spite of the differences, some similarities exist. For example, almost all legislation limits specific exposures (for example to heavy metals, benzene and benzidine). The lists differ in the chemical substances under control, as well as whether MAC values are accepted as criteria or not. The regulations may be revized on the basis of new research data.

This type of legislation faces many difficulties in view of the lack of experimental data on the role of gender in the reactions to the various substances. An even more difficult problem is that of individual vulnerability. No legislation deals with this problem with regard to the genotype; when MAC values are determined, a significant safety factor is usually used which takes into account both sex and individual differences. However, for some genetically hypersensitive workers MAC values are not a satisfactory protection. In such cases more radical measures are necessary, including removal of the worker from continuing exposure.

REFERENCES

Beliles, R.P. (1972) The influence of pregnancy on acute toxicity of various compounds in mice. *Tocixol. Appl. Pharmacol.*, **23**, 537–540.

de Bruin, A. (1976) *Biochemical Toxicology of Environmental Agents*, Elsevier/North Holland, Amsterdam.

Calabrese, E.J. (1973) *Toxic Susceptibility. Male/Female Differences*, Wiley Interscience, New York.

Krassovski, G.N. (1973) Extrapolation of experimental data from animals to man. *Health Perspect.*, **3**, 30–1.

Nikiforov, B. and Tabakova, S. (1981) Biological peculiarities of organism and toxic effect, in *Hygienic Toxicology, General Part.*, (eds F. Kaloyanova *et al.*) Medicine and Fisculture, Sofia, pp 44–59 (in Bulgarian).

Osobotná úprava pracovnysch podmienik zien (1972) Práca, Vydavatelstvo a nakladatelstvo ROH, Bratislava.

Panova, Z. (1988) *Occupational Obstetrical Pathology*, Medicina i Fizkultura, Sofia, P. 143 (in Bulgarian).

Raxman J.M. and Zuckermann, R.J. (1987) *Laws and Policies affecting Adolescent Health*, World Health Organization, Geneva.

Rozpordzazenie ministra pracy, prac i spraw socialnich z dnia 16 lipca (1979) w sprawie zatrudniania mlodocianych, który nim ukónczyli szkoly podstawowej. (Regulation of the Ministry of Labour and Social Affairs concerning the problems of young people without secondary education) *Dziennik Ustaw*, **16**, 232.

Rozporzadzenie Rady Ministrow z dnia 5 wrzésnia (1984) zmieniajace rozporzadzenie w sprawie wykazy prac wzbrinionych kobietom. (Additional legislation concerning the list of the kinds of occupation forbidden for women) *Dziennik Ustaw*, **44**, 548.

Tolokontsev, N.A. and Filov V.A. (eds) (1976) *Bases of the General Industrial Toxicology*, Medicina, Leningrad, pp. 213–18 (in Russian).

USSR State Committee for Labour and Social Affairs (1989). List of arduous and dangerous jobs, occupations and employment forbidden to young persons under 18 years of age. *Social Labour Bull.* **2** 208 (in Russian).

WHO (1987) *Children at Work: Special Health Risks*, Technical Report Series, No. 756, World Health Organization, Geneva.

Wykaz prac wzbronionych kobietom (1979) (List of occupations forbidden for women.) *Dziennik Ustaw*, **4**, 39–42.

Zielhuis, R.A. and Wibowo, A. (1989) Standard setting in occupational health: philosophical issues. *Am. J. Ind. Med*, **16**, 569–98.

Appendix: List of participants

GERMANY

Dr A. Basler
Federal Ministry for the Environment, Nature Conservation and Nuclear Safety
Bonn

Dr T.O. Goecke
Institute of Human Genetics and Anthropology
University of Düsseldorf
Düsseldorf

Professor G. Röhrborn
(Vice-chairman)
Institute of Human Genetics and Anthropology
University of Düsseldorf
Düsseldorf

TEMPORARY ADVISERS

Professor D. P. Agarwal
Institute of Human Genetics
University of Hamburg
Hamburg, Germany

Professor K. Berg
Institute of Medical Genetics
University of Oslo
Oslo, Norway

Dr N. P. Bochkov
National Research Centre of Medical Genetics
Academy of Medical Sciences
Moscow, USSR

Dr D. Djuric
Institute of Occupational and Radiological Health
Belgrade, Yugoslavia

Professor F. Kaloyanova
Department of Toxicology
Institute of Hygiene and Occupational Health
Sofia, Bulgaria

Dr S. R. Kleeberger
Department of Environmental Health Sciences
The Johns Hopkins Medical Institutions
Baltimore MD
USA

Professor I. Roots
Institut für Klinische Pharmakologie der Freien Universität
Berlin, Germany

Professor Philippe Grandjean
(Rapporteur)
Department of Environmental
Medicine
Odense University
Odense, Denmark

Professor J. Z. Hanke
Institute of Occupational Medicine
Lodz, Poland

Professor K. Hemminki
Department of Industrial Hygiene
and Toxicology
Institute of Occupational Health
Helsinki, Finland

Professor J. R. Idle (Chairman)
Department of Pharmacological
Sciences
University of Newcastle
upon Tyne
Newcastle upon Tyne, UK

Professor W. Kalow
Department of Pharmacology
University of Toronto
Toronto, Ontario
Canada

Professor H. W. Rüdiger
Unit of Toxicogenetics
Institute of Occupational Health
University of Hamburg
Hamburg, Germany

Dr V. Simeon
Institute for Medical Research and
Occupational Health
University of Zagreb
Zagreb, Yugoslavia

Dr. C. Sonich-Mullin
Environmental Criteria and
Assessment Office
US Environmental Protection
Agency
Cincinnati, OH
USA

Professor D. T. Teitelbaum
Medical Toxicology Partnership
Denver, CO
USA

Dr G. Winneke
Abteilungsleiter
Medizinisches Institut für
Umwelthygiene an der Universität
Düsseldorf
Düsseldorf, Germany

Invited but unable to participate

Professor E. J. Calabrese
Northeast Regional Environmental
Public Health Center
Environmental Health Sciences Program
School of Public Health
University of Massachusetts
Amherst MA
USA

Dr M. A. Mehlman
Department of Environmental
and Community Medicine
Robert Wood Johnson Medical School
Piscataway NJ
USA

REPRESENTATIVES OF OTHER ORGANIZATIONS

Commission of the European Communities (CEC)

Mr M. Curtis
Health and Safety Directorate
Luxembourg

Council for International Organizations of Medical Sciences (CIOMS)

Dr Z. Bankowski
c/o WHO
Geneva
Switzerland

WORLD HEALTH ORGANIZATION

Regional Office for Europe

Dr D. Kello
Toxicology and Food Safety

Dr S. Tarkowski
Environment and Health

Ms P. A. Christensen
Toxicology and Food Safety

Headquarters

Dr V. Bulyzhenkov
Hereditary Diseases Programme

Dr B. Chen
International Programme on Chemical Safety

Index

All references in **bold** type represent tables and those in *italics* are figures

MIX
Papier aus verantwortungsvollen Quellen
Paper from responsible sources
FSC® C105338

If you have any concerns about our products,
you can contact us on
ProductSafety@springernature.com

In case Publisher is established outside the EU,
the EU authorized representative is:
Springer Nature Customer Service Center GmbH
Europaplatz 3, 69115 Heidelberg, Germany

Printed by Libri Plureos GmbH
in Hamburg, Germany